Essential Matters

A History of the Cryptographic Branch of the People's Army of Viet-Nam, 1945–1975

with a supplement

on

Cryptography in the Border Guard
(formerly the Armed Public Security Forces)
1959–1989

UNITED STATES CRYPTOLOGIC HISTORY
SPECIAL SERIES, NUMBER 5

Translated and Edited

by

DAVID W. GADDY

Center for Cryptologic History
National Security Agency
1994

Foreword

The former Democratic Republic of Viet Nam (DRV – now the Socialist Republic of Viet Nam) emerged from revolutionary conspiracy, with roots in native independence movements as well as international communism. Ho Chi Minh – the name itself an alias – led what was originally a small band of revolutionaries in the period between the two world wars. Operating during the Japanese occupation of the 1940s as an underground resistance group (with incidental support from America's OSS) and continuing through the thirty-year struggle to establish the independence of the Democratic Republic of Viet Nam, first against the French, then the noncommunist Republic of Viet Nam, and, finally, the United States and its allies, his followers made secrecy a way of life. The leadership had noms de guerre (Vo Nguyen Giap, for example, was VAN). Given that background, the universal military penchant for abbreviations, acronyms, and nicknames, and a Sino-Vietnamese literary tradition that admired obscurity or hidden meaning, they produced an extraordinarily rich manner of expression, intended only for the initiate. Added to that was cryptography – secret writing – the art and science of codes and ciphers.

Essential Matters is a translation of a 1990 Vietnamese publication, *History of the Cryptographic Branch of the People's Army of Viet Nam, 1945-1975* (Hanoi: People's Army Publishing House). A supplement drawn from the *History of the Cryptographic Branch of the Border Guard, 1959-1989* (Hanoi: The Staff, Border Guard HQ, 1989), an organization originally known as the Armed Public Security Forces, extends the coverage by fourteen years, into the cipher machine era, and provides a natural complement. The Vietnamese of the titles (*co yeu*) literally means vital, essential, important, matters, with a heavy overtone of confidentiality or secrecy, as opposed to the normal Vietnamese word for cryptography, *mat ma*--the original name of this element of the People's Army of Viet Nam (PAVN), once known in the West as the Viet Minh army. (The American reader must bear in mind that "army," as used in the title, PAVN, more closely corresponds to "armed forces" in the United States, for air and sea components are subsumed in the term.) As a result of a decision made at the Eighth Army-wide Cryptographic Conference in February 1951, and in light of the international character assumed by Vietnamese interests in Laos and Cambodia, the euphemism was adopted by the security-conscious Vietnamese. But it was a euphemism that contained within it a reminder of the virtual mania for secrecy that elevated Vietnamese cryptography to such a critical role.

In this translation, both *mat ma* and *co yeu* are rendered as "cryptography." Personnel engaged in this pursuit are presented as "cryptographers," "cryptographic personnel," or even by the less elegant term, "cryppies." Women, as well as men, performed the function. Their organizations are rendered as "cryptographic," "cryptography," or simply "crypto."

Because of a century of French colonial domination in which the Vietnamese language was displaced by French, the generation of revolutionaries – or, at least, the younger ones they recruited--had little scientific knowledge of their own language, especially as a basis for applying cryptography. Learning as they went, by trial and error and precious few publications, they developed an indigenous cryptography, until, by the early 1950s, men trained in China returned to share the benefits of their learning. Presumably under Chinese influence, the native cipher systems were gradually superseded by enciphered codes. Always conscious of enemy cryptanalytic probing, improvement in maintaining secrecy of message content was a driving concern, adjusting for the educational level of their personnel and the circumstances in which they found themselves, with respect to geography, climate, and equipment.

The evident willingness of Hanoi authorities to break their traditional silence and permit the public reading of this work may be interpreted in several ways: perhaps the rigid protection of all aspects of Vietnamese cryptography has abated, at least for the period covered by these two books, a period that ended nearly two decades ago. Perhaps the techniques described are now passé. And in any event, as time moves on, the anonymous crypto-warriors age and die: for many, this was the last opportunity to see their names in print, and to share in the telling of their unsung role in winning the final victory. In any event, the result is an extraordinary contribution to the history and literature of cryptography. It affords insight into a previously hidden aspect of the military life of a people who, for a few years, held the center stage in world attention. With the passing of that generation, it preserves the names of participants, men and women who once led lives of intentional and enforced obscurity. It tells us of their training and their accomplishments, their hardships and suffering. It tells of the toll they paid – some 500 cryppies paid the supreme sacrifice, nearly 10 percent of those on duty as of 1972. It forms the tradition for the coming generations.

These men and women had created an effective communication security system literally "from scratch." More conversant with French than their native language (which now represented nationalist aspirations), they had to subject their language to the most basic analysis of structure, its specialized military and technical vocabulary, the frequencies of its letters and words, and its rendering for cryptographic and radio transmission purposes. As in other aspects of Vietnamese military life, deprivation was made a virtue: lacking the ability to establish central control over the production and use of cryptomaterials; standards were set and models adopted; then local initiative was encouraged through competition and emulation campaigns. An enemy was thus confronted, not with an Enigma or "Purple" to break, so much as a wide variety of similar cryptosystems, having to be attacked individually, much as was the challenge offered the Allies by Japanese army systems in World War II. (An interesting and instructive companion to the present translation can be found in Edward J. Drea, *MacArthur's ULTRA: Codebreaking and the War Against Japan, 1942–1945* [University Press of Kansas, 1992], which also illustrates the principles of enciphered code.) Production figures, tonnage of materials delivered, and scores attained in cryptographic competition ("Comrade Nguyen Van Hai encrypted x groups in y minutes with an error rate of only z") seem tedious or even silly to the Western reader, until one realizes the importance of such matters in an army using often crudely printed (or even handwritten) manual cryptosystems, distributed over rugged, distant jungle trails by couriers or man-pack, later to be transmitted by finger or tongue.

In a form reminiscent of the old revolutionary army, lacking ranks and titles, other than "warriors" and "cadre," and with individuality subordinated to "the team," the authorship of this work is identified collectively. The senior of the two officers identified as being responsible for its content, Brigadier General Nguyen Chanh Can, was a graduate of the first formal Vietnamese class in cryptography, September-October 1946, selected to remain at the General Staff Cryptographic Bureau. He may also have been one of the original forty-five-man Vietnamese group sent to China for some six months of training, returning in May 1951, when, as bureau deputy chief, he was concurrently made chief of the Campaign Cryptographic Section of the reorganized Cryptographic Bureau of the General Staff. He headed the Cryptographic Section of the historic Dien Bien Phu campaign (1953-1954), and he figures elsewhere in the text. We can thus assume that the book represents the efforts of both participants and a later generation of researchers, editors, and publishing staff, as presumably is the case with the book drawn upon as a supplement.

At the same time, the book lacks the detailed documentation expected in comparable Western military histories, leaving one to wonder to what extent documentation has been preserved and the extent to which recollections play a major role in this account, making it the basic documentation for the future and, in the process, shaping the traditions and perceptions of coming generations in the speciality.

Finally, and in a departure from conventional orthography, this translation has rendered the Vietnamese letter, "unbarred D," as "Dz," approximating its sound, by contrast with the "barred D," comparable to the "d" sound in English. (This results, for example, in the name of General Van Tien Dung being rendered as Van Tien Dzung, avoiding the unfortunate American tendency to call him "General Dung.")

DAVID W. GADDY

Notes on the Translation

By comparison with the extensive vocabulary of English, its synonyms and its free borrowings from other languages, Vietnamese has a more limited vocabulary to draw upon, but a vocabulary filled with nuance, heightened in the case of communist usage. Strict consistency in translation, while desirable in one sense, would produce a highly repetitive text to the English reader; therefore no effort has been made to ensure consistency when American English seems "richer." The risk is, of course, misinterpretation. An attempt has been made to render certain Vietnamese terminology in equivalent American English, but to avoid "forcing" an interpretation if the precise American equivalent was unknown. This is especially true of technical cryptographic terminology, when a forced meaning could well be a misinterpretation. For example, in normal usage, *tai lieu* would be translated as document or (written) material, but in a cryptographic context, "cryptomaterial" seems warranted. *Luat*, code in Vietnamese, is sometimes rendered as "code," at others as "system," or "cryptosystem," depending upon context.

Certain Vietnamese terminology carries the connotation of echelon (e.g., *tieu ban, ban, phong, cuc, tong cuc*), requiring consistency in translation. Thus the change of a *ban* (section) to a *phong* (bureau) or a *phong* to a *cuc* (directorate) implies bureaucratic growth --elevation and a promotion for the chief. Used in a political or governmental context, *ban* is translated as section. In other instances, inconsistency may seem the rule: "Branch" may be Vietnamese *nganh* or, in military terms, *binh chung*. "Sector," "zone," and "region" may involve the Vietnamese words, *khu, khu vuc, xu* (in the 1940s), *mien* or *vung*. In such cases--and, in general, whenever the Vietnamese appears necessary for comprehension--the original term is given in brackets, as are other interpolations or comments. (By contrast, parentheses are used only as in the original text.)

The book often uses the ellipsis (. . .) in a manner more akin to "etc." or "et al." in our usage. To avoid the misimpression of omission in the translation, "etc." is usually substituted for that practice.

The title "comrade" is so pervasive that it has been abbreviated as "Cde" in the translation. Other abbreviations are standard (e.g., HQ for headquarters, CP for command post, MR for military region), but are spelled out in first instances.

Terminology associated with party and military organizations may also require explanation. "Central" or "Central Party" might well have been rendered in John Le Carré fashion as "The Center," but this has become an accepted way of rendering *Trung uong* or *Trung uong Dang* in English, referring to the headquarters of the Lao Dong Party – or frequently, "the top," in light of the intertwined party-government-military structure French authorities called "parallel hierarchies." Strict consistency is the rule for military organizations, as noted earlier – *Bo tu lenh* is "headquarters," at division, military region, or branch/service level; *Bo chi huy* (an earlier term) and *bo tong tu lenh* are both rendered as High Command (GHQ, in some usage); and *Bo tong tham muu* as General Staff. *"Bo"* by itself is often used, meaning "the top" for the military structure: is this the High Command or the General Staff? Or might it be the Ministry of National Defense (*Bo Quoc phong*)? Here *bo* is rendered as simply "Headquarters" or "HQ." "Division" has been used to translate both the *dai doan* of the First Indochina War and the postwar *su doan*, the product of "modernization and regularization." On the other hand, the somewhat artificial term, "groupment," has been used for *binh doan*, the forerunner of *quan doan*, "corps," for the Vietnamese used the same term, *binh doan*, for the French "mobile groups" of the First Indochina War, and "group" is the consistent rendering of the Vietnamese *doan*. Military rank tends to be consistent with "normal" American translations, with the exception of Senior General, which is rendered simply as General.

There is a long, if not necessarily honorable, history behind these inconsistencies in English: for some reason, we have always accepted Ho's title as President, notwithstanding the fact that the same term is rendered as Chairman in the case of Mao. Of course, it is rather as Uncle that we find him in the following pages.

DWG

History of the Cryptographic Branch of the
People's Army of Viet Nam
1945 – 1975

**The Publishing House invites the
opinions and criticisms of the readers.**

LỊCH SỬ
NGÀNH CƠ YẾU
*QUÂN ĐỘI NHÂN DÂN
VIỆT NAM
(1945 - 1975)*

*NHÀ XUẤT BẢN QUÂN ĐỘI NHÂN DÂN
Hà Nội - 1990*

Facsimile of original title page

The History of the PAVN Cryptographic Branch is written to cover the building and activities of the army cryptographic branch from the resistance against French colonialism through the national salvation opposition to America (1945–1975).

Content direction: Brig. Gen. NGUYEN CHANH CAN
                     Col. PHAM VAN THIEU

Research and Compilation: LE DINH Y
                            HOANG QUYEN
                            VU CONG SUU
                            VU VAN TAN

Manuscript: VU CONG SUU

Ho Chi Minh – "Uncle Ho"

"Cryptography must be secret, swift, and accurate.
Cryptographers must be security conscious and of one mind."

Words of Uncle Ho, with the cadre and students
of the 1950 Viet Bac Combat Sector army
cryptographic class

"During the decades past, as we fought the aggressor, we knew that, at the most crucial times, our most secret matters would not be leaked out.

"You comrades have participated most importantly in maintaining secrecy, contributing to our common strength in achieving victory.

"The cryptographic branch – a most important branch – must bring itself fully up to date; the ranks must be truly pure; the regulation of the task must be very tight."

From the speech by Cde Le Dzuan,
General Secretary, Central Party
Executive Committee, at the 1978
Nationwide Cryptographic Cadre Conference

Comrade Le Dzuan, General Secretary of the Central Party Executive Committee, speaking at the Army-wide Cryptographic Cadre Conference, 1978

**General Van Tien Dzung, Central Party Politburo member and
Minister of National Defense, checks out research results in the use of cipher machines (1984)**

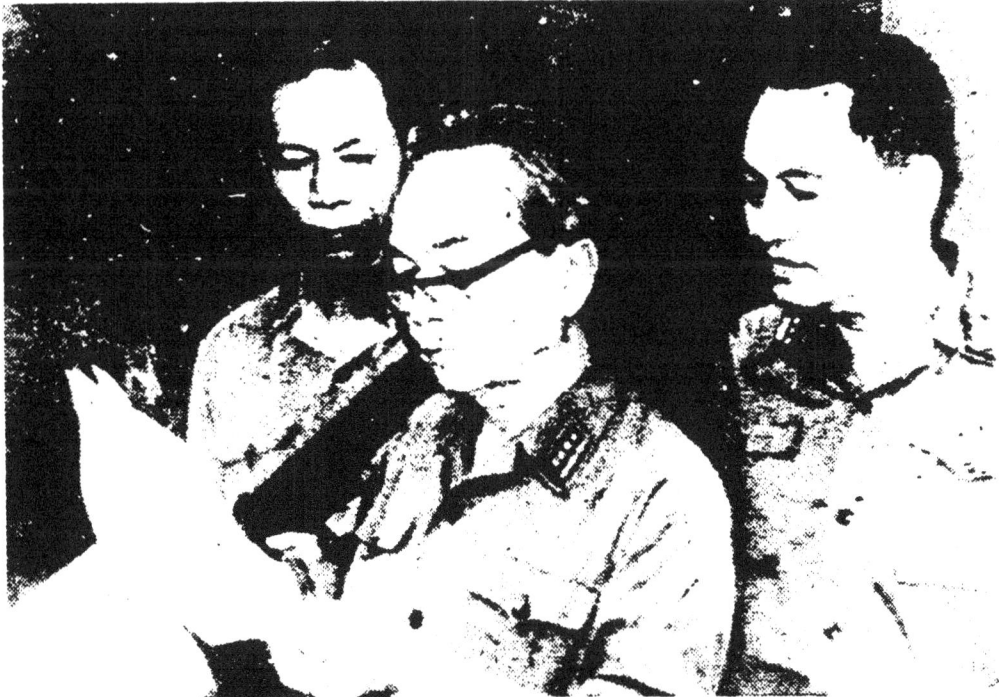

**General Vo Nguyen Giap, Member of the Central Party Politburo and Minister of National Defense,
reads a report message from the front sent back during the Spring 1975 general offensive and uprising**

Words of commendation and recollection to the crypto cadre and personnel from General Vo Nguyen Giap, Minister of National Defense

Table of Contents

[Transposed from the end of the original volume]

Chapter One

The Genesis of the Cryptographic Branch
of the People's Army of Viet Nam

APPEARANCE OF THE EARLIEST CRYPTOGRAPHIC ORGANIZATIONS
AND TECHNIQUES IN THE ARMY

The August Revolution was a success!

On 2 August 1945, in Ba Dinh Square, President Ho Chi Minh solemnly read the Declaration of Independence, giving birth to the Democratic Republic of Viet Nam.

The newly established democratic republican government had to cope with a situation of endless complications. In the South, the French army, hiding behind the British military, landed in Saigon-Cholon, continuing their aggressive plots against our nation. In the North, nearly 200,000 of Chiang Kai-Shek's military came, in the name of the Allied Forces, to disarm the fascist Japanese, and escorted gangs of lackies plotting to overthrow the authority organized by the Viet Minh [Viet Nam Independence League]. Reactionary gangs in the nation took advantage of the people's rising to act to counter and destroy the resistance. The Japanese-French imperialist gangs provoked famine and ran wild.

Faced by the devious plots of enemies both foreign and domestic, together with difficulties in every respect in our homeland after the revolution had just succeeded, the fate of our nation at that moment faced a very dangerous situation, for "money and necessities were hanging on a hair."

In order to confront the enemy's aggressive plots against our nation--in order to protect the incipient authority of the people--the Standing Committee of the Central Party and President Ho Chi Minh proposed some urgent tasks to strengthen authority, counter the French colonialists' aggression, abolish crime, raise the standard of living of the people, and especially to show concern for supplying concrete guidance in building revolutionary armed forces.

On 7 September 1945, President Ho Chi Minh entrusted to Cde Hoang Van Thai responsibility for establishing the General Staff organization. (Cde Vo Nguyen Giap was also present at the occasion). When assigning the task, Uncle said: "Our people have just won their independence, their freedom--all of our nation is starting to build an army of resistance and self-defense to safeguard our independence, our freedom. Pursuant to

1

collective instructions, the Staff [Bo tham muu] is established to help Central exercise command of the army in our whole nation.

"As the secret organization of the collective--as the nerve center of the army-- the Staff is responsible for military strength, for forging weaponry, for knowing the enemy, in order to defeat every enemy.

"At this time we have no experience--do not yet understand staff work--have many difficulties. But we have to strive to overcome this, studying even as we work. With determination, many difficulties can be worked through. Somehow we must also build a staff branch of our army that is solid and powerful, worthy of the tradition of giving one's mind to forwarding and preserving the nation of the Vietnamese peoples."[1]

On 8 September 1945, Cde Hoang Van Thai sponsored the first meeting of comrades to introduce the organizations of the General Staff, in order to determine and allocate responsibilities.

On 9 September 1945, the Military Communications-Liaison Bureau of the General Staff was officially established, under Cde Hoang Dao Thuy. From a week before, according to instructions from the Central Party Standing Committee and Uncle Ho, Cde Vo Nguyen Giap, Minister of Internal Affairs in the provisional revolutionary government, and concurrently Commander-in-Chief [Tong chi huy] of the revolutionary armed forces, had a personal exchange with Cde Hoang Dao Thuy and handed to him responsibility for preparing to build a military communications-liaison system. This was a responsibility that had to be carried out at once, with no delay, in order to help Central, the government, and the High Command [Bo Tong chi huy] come to grips with the situation and issue timely instructions to the combat sectors [chien khu] and units throughout the nation.

Immediately upon being established, the Communications-Liaison Bureau started to develop the organization of a network [mang luoi] of communications-liaison, and the first task was the early creation of a radio liaison network especially for the army.

With the Central Party concerned with assignment of people and getting equipment, a few of the fellows took it upon themselves to search out old radio stations in French depots, and, after just a short time, an official military radio liaison net was inaugurated, comprising the High Command and General Staff in Hanoi with the combat sectors of Dong Trieu, Viet Bac [Northern Viet Nam, or Tonkin], and Hoa (Binh)-Ninh (Binh)-Thanh (Hoa); the Revolutionary Military Affairs committee of Trung Bo [Central Viet Nam, or Annam]; and the Military Affairs Committees of Thua Thien-Hue and Da Nang. Besides the military liaison network, liaison between the Central radio offices (Hanoi) and those of Trung Bo and Nam Bo [Southern Viet Nam, or Cochin-China] was maintained and shaped, making a liaison network of military postal and radio [service] throughout the nation.

In the first days of the revolutionary regime, the leadership and command comrades, the organizations sending messages and those receiving messages, and the

2

communications organizations equally felt uncomfortable sending messages over the military radio or via the post offices in plain text, unenciphered. Thus an urgent requirement was to research methods of using cryptography so as to ensure communication security.

Cde Hoang Van Thai personally reviewed and approved the plan to establish the Cryptographic Section [Ban Mat Ma] (not yet called Co Yeu, as it was later) and strongly recommended that Cde Hoang Dao Thuy select good, trustworthy, literate people for the task of cryptography.

On 12 September 1945, the Military Cryptographic Section, the first cryptographic organization of the army, was established, tasked with research, production, and use of cryptographic systems to ensure the secrecy of leadership and command communication of the various echelons of the army going via the various means of communication (principally by radio). This was also the first cryptographic organization of our nation.

The twelfth of September has since been taken as the birthday of the army cryptographic branch.

At the beginning, the Cryptographic Section lay in the Bureau of Communication-Liaison. The working area was a room behind that of Cde Bureau Chief Hoang Dao Thuy in building No. 16 Riquer Street,[2] adjacent to the General Staff organizations.

In accordance with a proposal by Cde Hoang Dao Thuy, Cde Vo Nguyen Giap appointed Cde Ta Quang De,[3] who was working in the Ministry of Internal Affairs, to transfer to the General Staff and assume charge of the cryptographic mission.

Having received the responsibility and concrete recommendations of Cde Hoang Van Thai, and after determining the particulars and mission components, Cde Ta Quang De circumspectly proceeded to select people who could be introduced to the work of cryptography. The "criterion" for selection was based on estimation of a "good, trustworthy person," and an additional condition was "ready-to-go," no family ties, committed to an assignment requiring total reliability. A number of the fellows who were introduced by responsible people to cryptographic work prior to the organization were intellectuals and petty officials, and at the same time Boy Scouts from [troops] such as Dinh Loan. Thuyen, Hoang Quy Quan,[4] also young Miss Bui Thi Loan, a liberation army soldier returned from the combat sector. After a few days, Nguyen Tu Khang, Bach Tuong, Sam, Dich, then Tran Mi Thach in turn were introduced to the work.

After working together a few weeks, the fellows did some brainstorming about many aspects of this new assignment. Prior to this time, the colonialist French gang had not ever instructed Vietnamese people, or entrusted them, in our own cryptographic organizations, so there was no one who thoroughly understood the business. Therefore, the ideas shared by these fellows, together with revolutionary enthusiasm, built up trust on the part of the leadership comrades. Afterwards, in a situation in which people to perform cryptography were lacking in various places and in accordance with arrangements by Cde Hoang Van Thai, Dich went to Combat Sector 4, Thach went up to

Thai Nguyen-Bac Can, and Sam, Khang, and Tuong were placed on a long line from Phuc Yen up to Phu Tho, locales that needed to quickly put down the activities of the [Vietnamese] Nationalist Party and "Viet Cach" [Vietnam Revolutionary] cliques.

Near the end of the year, the Cryptographic Section received two more people: Cde Nguyen Hai Hac, graduate of the Higher Agricultural School, home in Hanoi, and Cde Tung Anh[5] from Quang Ngai, who came at the invitation of the military organization there.

As of this point, people working in the Cryptographic Section were temporaries. Some arriving before, some later, these were the comrades present in the early stage of the formation of the army cryptographic branch. A small, close-knit and affectionate family, united to help each other, with a clear sense of responsibility and personal honor to respond to the requirements of the revolution and the army, these comrades shouldered a tough, essential job, doing the spade work for a technical branch of the PAVN.

Initial feelings of inadequacy passed quickly. With respect to cryptography, it could be said that the resources of the comrades at this stage were simply revolutionary zeal and enthusiasm, together with a book, *The Basic Principles of Cryptography*[6] in French, which had come apart, and some two or three Boy Scout riddle games[7] – this was the level of knowledge at this stage.*

* A variation in the account of the origins of Vietnamese cryptography is contained in Volume 1, 17-18, of the *History of the Communications-Liaison Troops* (Draft) (Hanoi: Communications-Liaison HQ, 1985): "[In the fall of 1945] when they sent out an unenciphered official message, all of the Communications-Liaison Bureau saw it and were uncomfortable. The place that received it also asked why it wasn't encrypted. . . . The chief of the Communications-Liaison Bureau proposed to HQ the establishment of a Cryptographic Section (not yet called 'essential matters'), to be placed in the Communications-Liaison Bureau. Imagine that! Cryptography, placed in the Communications-Liaison Bureau! But that is the truth. We need to say more about this fresh new organization: Beforehand, throughout all of Indochina, the gang of French colonialist rulers would not train and entrust to the Vietnamese employment in their cryptographic organizations. That's understandable. As a result, when we seized power, the two radio liaison centers, north and south, had to generate two reference works for enciphering and deciphering that differed from each other. In Hanoi, when the first messages were sent to Saigon, they requested the recipients to use passages in the classic, Kieu, by Nguyen Dzu as an enciphering-deciphering reference work. In Saigon, they proposed that Hanoi use Mendeleyev's periodic table in order to solve the key and encipher and decipher the messages sent from then on. Thus, it was imperative to have a [or, "one"] cryptographic organization. Not knowing where else to put it, it was properly placed in the Communications-Liaison Bureau.

"On 12 September [1945], the Cryptographic Section was established, subordinate to the Communications-Liaison Bureau.

"Cde Ta Quang De (Ta Quang Dam) and Cde Dinh Loan Thuyen (Hoang Thanh) were invited by the General Staff to undertake the making of cryptosystems. Before the August Revolution, in the scientific games and entertainments of the students and intellectuals in Hanoi, there were many young people--among them Cdes De and Thuyen--who regularly played number and letter games according to set rules, and transformed the numbers and letters to make words and sentences. Now, confronted by the requirements of the revolution--of the army--the comrades readily accepted and got involved with a deep attitude of responsibility. One week later, a cryptographic paper with adequate key was accepted by the General Staff, and the Communications-Liaison Bureau was instructed to quickly develop cryptographic personnel for each radio station."

Ibid., 20: ". . . the troop strength of the Bureau [at this time] (including the bureau chief) was only eight people. . ." [footnote 2:] "8 people, comprising Hoang Dao Thuy, Le Dzung, Vu Han Thang, Ta Quang De, Dinh Loan Thuyen, Nguyen Ai Hac, Do Thanh, and Tran My Thach." -- Tr./Ed.

The comrades divided among themselves going to meet comrades who had experience in covert activities inside and outside our country, seeking to learn from their experience; searching for a few types of [cipher] keys, easy to remember, easy to use; a few methods of writing secret letters; all sorts of text books in the Vietnamese language that could be consulted in research. The comrades regularly said to one another, "Now we've really got to get back to our mother tongue!" Not long before this, every time one spoke up in class it had to be in the language of the French "mother country," and now, how eagerly they were going to work in the Vietnamese tongue! There were young men and women at this time, under the age of twenty, coming to grasp the fact that Vietnamese had many alphabets and knowing which, encountered many, especially in literature. The [cryptographic] concept of "frequency" emerged through the practice of enciphering and deciphering.

At the suggestion and urging of Cde Hoang Dao Thuy, after about a week, Cde Ta Quang De researched a system to use to write and read secret letters.[8] Per instructions from Cde Hoang Van Thai, this cryptographic system was given to a command comrade on the Vinh Yen front to use for liaison. Some days later, Cde Hoang Van Thai gave Cde De a secret letter to encipher for sending to the Vinh Yen front. The content was a short section, but written out to produce a full page. When signing the message form, the comrade chief of the General Staff observed, "First, this way you have to write a lot, use up a lot of time, a lot of paper; second, it's rather difficult to ensure a fit between what's valid and what's false, thus the end is revealed; and, third, it's not handy for message transmission."

From these observations and exchanges with the fellows in the Cryptographic Section, Cde De came up with a different method: a method of enciphering each alphabetical letter--simple in use, a neat, compact chart. In using these systems for monoalphabetic [doc bieu] substitution and transposition, by close observation of a number of cryptograms, the comrades subjected them to detailed scrutiny and saw a number of points arise: Concerning monoalphabetic substitution: the repetition of the plain-cipher values followed a basic form. Some consonants and vowels in the Vietnamese language, represented by two letters, caused the frequency to rise very clearly.

Enciphering by transposition, using a literal key or a digital key: If enciphering by simple letter transposition, then the number of letters of the enciphered message would be equal to the number of letters of the plaintext message. These conditions resulted in cryptography that would not ensure tight secrecy, easily exposed to the eye of an enemy with much experience in cryptanalysis. Each individual arrived at the conclusion that the systems being used would not yet meet the requirement to ensure secrecy, their degree of security was at a rather low level – we had to quickly research different systems to replace them.

Based on experience with monoalphabetic substitution cipher, Cde Ta Quang De discussed with the people in the Section finding a way to change the system and change the key. Above all else, to produce some way that the appearance of the cryptogram was

5

not dependent upon the appearance of the plaintext message--when our people looked at the cipher text they only saw it in its entirety, whereas clusters of letters differed from others in frequency and quantity. Then, in enciphering, to carry out a key change, a chart or direction change, even right in a message. The concrete method would have to have a composite chart, composed of many contrasting systems, each system with an individual key. After a period of searching and exchanging views within the Section, Cde De came up with a new type of system. The first columnar chart system had succeeded, with Cdes Hoang Van Thai and Hoang Dao Thuy agreeing to introduce its use to replace the systems that had been used for so long. Once the cryptographic systems had been typed up, Cde Hoang Van Thai sent a message to the sector chiefs to select dependable people to come to HQ to receive the system and listen to directions for use. A few days later, Van Tien Dzung came from Chi Ne bringing along a cryptographic cadre, then Le Van Suu from Intersector 4, Vu Hien from Haiphong, Le Quang Hoa, Hoang Minh Thao, et al., and the other units in turn selected cryptographers from their units to receive the system.

According to the principles of this system, each cryptogram had to be changed into a fixed number of columns. Each column had its own key and in order to be certain of ensuring secrecy, each key was fixed for many cryptographic values [ky hieu mat]. At first, many comrades were a bit perplexed, suggesting that use of simple transposition would also ensure secrecy. The comrades in the section had to stand their ground in explaining, and the others finally became truly unanimous. After a period of use, enciphering and deciphering many times, the work became routine. The comrades in the research team continued to realize an additional step, regulating the movement of the cipher strip [viec chuyen bang ma] according to class [theo bac], according to day, etc.

One might say, from forms of "leave out the false, what's left is valid" in order to read secret letters (not yet forms that could properly be called cryptosystems) and monoalphabetic substitution and transposition by pattern, we quickly moved to chart systems using irregularly arranged values with many columns and many keys. These chart systems, given the educational and technical conditions of our society, generally speaking, and our army, specifically, at that time, seemed relatively suitable and achieved a notable level of meeting the requirement to serve guidance and command.

The work came fast and furious, the activity of the Vietnamese Nationalist Party in the provinces of Phu Tho and old Vinh Phuc being reported continuously by message. Then the Nam Bo resistance, the French colonialists hiding behind the British military, swarming into Indochina and attacking and occupying many cities. The endless struggles of great fortitude of our Southern compatriots took place daily. So as to quickly grasp the situation and issue guidance for coping with the enemy and foreign aggressors, the Cryptographic Section, along with the communications-liaison organizations, worked day and night. From the end of September, through October – November 1945, the volume of secret messages increased very rapidly. The matter of telegraphic language frequency posed a need for research into a method of scientific calculation. (Initially, this matter had to be done by guessing, for there was a time-sensitive need to satisfy the most pressing, urgent requirement for the service of command.) The clear part of the cryptographic system form received concentrated research in construction. Combining experience and research and the use of a columnar system that followed the principle of monoalphabetic substitution with a digital key, with personally searching out and consulting foreign documents in order to apply international principles suitable to the special aspects of Vietnamese telegraphic spelling, Cde Dinh Loan Thuyen and colleagues produced a Vietnamese 676-cell [26x26] chart. The system was constructed according to the method of polyalphabetic [da bieu] substitution encipherment of word components. After completion of the system, it was quickly put into use.

Implementing a directive from the Chief of the General Staff, the Cryptographic Section designated people to convey the system to a number of places lacking the means to receive it. On the first lunar new year's day after the August Revolution, Cde Thuyen was ordered in turn to convey the system and directions for use of the 676-cell chart to the units from Phu Ly to Binh Dinh. Vis-a-vis Nam Bo, that theater of war received concentrated assistance from the whole country: from October 1945, the Cryptographic Section quickly sent cipher models, carried by Cde Hoang Quoc Viet, to be handed over to the Sector [Xu] committee, when the comrade, on behalf of the Standing [Committee] of the Central Party, went down and participated in the opening at the Sector Committee Conference.

A consciousness of the need for constantly changing keys, changing cipher strips, replacing systems dawned early-on among the comrades engaged in cryptography.

One day, around the end of 1945, the Second Bureau [Military Intelligence] sent over to the Cryptographic Section a number of enemy cryptograms. Although knowing nothing about cryptanalysis, the fellows nevertheless took a stab at it. The unexpected result was that the comrades decrypted a third of the cipher messages from a French army unit stationed in Upper Laos exchanging operational matters among themselves by means of a simple substitution system. This resulted in the Second Bureau comrades forming a high opinion of the expertise of the cryptographers. As for us, this small accomplishment in cryptanalysis had the effect of helping our people make up systems, even as it helped them in encrypting messages themselves.

In January 1946, in accordance with instructions from Central, the General Staff organizations were realigned. The cryptographic organization was split off from the Communications-Liaison Bureau. For the political, routine, and administrative aspects, [it would be] directly subordinate to the General Staff secretariat; for speciality professional knowledge that was the essence of its task, directly subordinate to the comrade Chief of the General Staff. The main office of the bureau was moved to No. 26 Hang Bai (a sign in front of the door of the office said "Bureau of Secret Messages"). Crypto elements were also gradually established. In the sectors [khu], the cryptographic organization was called "Department [ty] of Secret Messages," or "Department of Cryptography." In Sector 1, the Department of Secret Messages was organized directly subordinate to the regional command post secretariat, while in the regiments cryptographic teams were directly subordinate to the regimental command section. In sectors 4 and 5, the cryptographic organizations were called

Department of Cryptography, directly subordinate to the secretariat or staff organizations, with cryptographic teams in the regiments.

In March 1946, Cde Ta Quang De became a liaison control officer between the Vietnamese and French sides.[9] Cde Hoang Van Dong became Chief of the Cryptographic Bureau. Research into cryptographic technique was undertaken by Cdes Dinh Loan Thuyen and Hoang Van Dong.

At this time, the 676-cell code chart was improved, fixed strips replaced by movable strips, raising the level of cryptographic security. The content of the chart was added to by compound words [tieng kep] and phrases [doan cau], thus shortening cipher messages. The system received use at once and was brought into play with effect in ensuring communication security between the High Command of the Vietnamese relief troops [Bo Chi huy Tiep phong quan Viet Nam] in Hanoi, with eleven units. Hai Dzuong, Thai Binh, Phu Ly, Nam Dinh, Ninh Binh, Thanh Hoa, Vinh, Dong Hoi, Dong Ha, Hue, and Da Nang dealing with the French army.

At the beginning of 1946, Cde Thuyen received the task of preparing a system to ensure cryptographic liaison between our government's delegation at the Da Lat preparatory conference and the General Staff. Realizing the degree of importance of the matter in our foreign relations struggle with the French, to ensure the secrecy of policy instruction laid down to our delegation by the party and government, Cde Thuyen concentrated all of his efforts into making a system. The good things of the chart form were incorporated into a chart system made solely to serve the Da Lat conference. The leadership comrades also knew that, at this conference, the French side had brought along a guy who was a specialist in cryptanalysis, with equipment to search out our secret information. Having received the first two messages sent back from the Da Lat conference, Cde Thuyen personally broke them out and was flabbergasted: the person enciphering the messages had not conformed closely to the regulations concerning the technique that had been conveyed--the information we were sending could easily be uncovered by the enemy. The Cryptographic Bureau at once suggested to Cde Hoang Van Thai suspending the use of this type of system. During the course of working for the previous half year, Cdes Dam, Thuyen, and Dong were tormented with mixed feelings, principally over whether the cryptography they were using was really tightly ensuring military secrecy or not. After reading some documents that had been received, the comrades knew that those people were still enciphering and deciphering by machine, and especially that they were still using equipment that made it very easy to discover the system and key. Thus they had to be even stricter with themselves, although believing firmly in the sense of loyal service by the young men and women performing cryptography, as far as their country was concerned, the general standard was still low and capacity for carelessness no small matter. The comrades discussed with the young men and women in the section a cryptanalytic test of an enciphered message, setting aside knowledge of system and key, to see if it could be made out. But, although they worked out the system and key, once, when they did not have system and key at hand, no one could make out anything at all. Thus day by day and week by week, continuing this sort of thing, they listened intently and watched for signs of any leaks in secrecy, and they felt reassured.

In April 1946, in order to increase the building and creation of conditions for the cryptographic organization to fulfill its mission in situations that had become more urgent daily, the cryptographic organization of the General Staff was supplemented by comrades Ho Ton Vinh, Hoang Don, Hoang Tuyen, Luong Dzan, Hoang Dzanh Cha--Cde Ho Ton Vinh (alias Hoang Duc Ton) was one of the first three Communist party members in the General Staff organization and was sent to the cryptographic organization as a supplementee. These young men were introduced by leadership and command comrades of high standing.

Along with the building and strengthening of the cryptographic organization at Central, the training of cryptographic cadre to expand the system of cryptography and strengthen the units was fully appreciated. As a matter of urgency, the General Staff cryptographic organization prepared to open a mass class to train cryptographic cadre to supplement the cryptographic organizations at lower echelons. The task of research and compilation of [cryptographic] material to teach in this class was given to Cdes Dinh Loan Thuyen, Hoang Van Dong, and Hoang Dzanh Cha. Although they encountered many difficulties, the comrades diligently consulted, researched, and compiled a quantity of documents concerning the science and techniques of cryptography.

Material compiled by Cdes Hoang Van Dong and Dinh Loan Thuyen first-off comprised basic theoretical content and instruction in the use of cryptographic systems. Cde Hoang Zanh Cha researched and compiled the statistical content concerning Vietnamese language frequency, in order to serve in constructing cryptographic systems. These were also the formative works in the sphere of research into the science of cryptographic technique on the part of the army cryptographic branch.

After preparing sufficient documentary content to cover all aspects, the General Staff decided in September 1946 to open the first class in the army to train cryptographic cadre: it was named "The Hoang Dzieu Class." Sessions took place at No. 7 On Nhu Hau Street (now Nguyen Gia Thieu Street), Hanoi, under Cdes Hoang Van Dong, Hoang Dzanh Cha, and Ho Ton Vinh. Cde Hoang Van Dong was in general charge and bore responsibility for teaching cryptographic principles. Cde Ho Ton Vinh was responsible for political leadership and thought and bore responsibility for teaching political matters. Cde Hoang Dzanh Cha bore responsibility for teaching the methods of researching the frequency of the Vietnamese language and the use of cryptographic systems. Twenty students from the combat sectors and units in Viet Bac, Trung Bo, and the Southern Viet Nam Resistance Committee (in Quang Ngai) were selected to attend the class. Nam Bo, hindered by transportation, was unable to send up people to attend. Before getting into specialized study, the students had to grasp the mission thoroughly, clearly define ideology, and understand the "must-do's and the musn't-do's" of the work of people performing the task of cryptography. Beyond the basic content of the syllabus, the class also received an additional introduction to fundamental knowledge of the cryptology of the world and practiced a number of our forms of cryptography. After more than a month of study, achieving good results, the class came to an end. The students returned to the units, becoming the nucleus of building cryptographic organizations in the combat sectors [chien khu]. Some comrades were retained to supplement the General Staff cryptographic organization, namely, Nguyen Chanh Can, Hoang Manh Tuan, Vu To, Vu Duc Minh, and Nguyen Van Dzanh (Ho Quang Chinh) – Cde Ngo Vi Thien was sent to supplement Combat Sector 1; Cde Tran Dac Quy to Combat Sector 2; Cde Le Hai to Combat Sector 3; Cde Van An to Combat Sector 4; Cde Dong Tam to Combat Sector 5; Cde Tung Anh to Combat Sector 6, etc.

In May 1946, in order to implement Ministry of National Defense instructions to organize reliable and secure communications-liaison nationwide, the cryptographic organization of the General Staff convened the first cryptographic conference. The conference discussed the tasks of cryptography and secrecy, and the delivery and receipt of cryptographic systems. Comrades in charge of the cryptographic organization in the combat sectors and fronts came to attend (except for cryptographic cadre-in-charge in Sector 5 and Nam Bo, who could not get back to attend). The conference received a visit and teaching from the Cde Chief of the General Staff, Hoang Van Thai. In August 1946, meeting again in the second military cryptographic conference in Hanoi, twenty-five delegates from Trung Ky [Annam] and Bac Ky [Tonkin] attended: the discussion was on cryptographic training. After the conference there was a five-day specialty training course.

Thus, together with resolute research, ingenuity, and expansion of the cryptographic net, having received the concern of the leadership and command cadre, the cryptographic organization urgently strove to train cryptographic cadre and personnel to supplement the units. From the beginning of December 1946, the cryptographic organization in the army had, by turns, organized in the combat sectors, the companies [chi doi], and a number of units, army-wide, participating in "maintaining contact between the combat sectors, an essential condition for unified command," in the spirit of "Chi thi khang chien kien quoc" ("Instructions for Resistance in Founding a Nation"), from the Executive Committee of the Central Party (25 November 1945).

Notwithstanding, during this time we had not yet come out with concrete regulations, so getting and using cryptographic cadre and personnel remained hit or miss. The cryptographic organizations in the sectors were not able to manage the total number of personnel under their authority. The command organizations routinely transferred cryptographic personnel to other assignments, or assigned them work outside the sphere of their technical speciality. This situation worked counter to the ideology of a number of cryptographic cadre and personnel, principally at the regimental and company level. A number of cryptographic personnel had no stomach for the job, and requested direct combat or transfer to some other duty.

SERVING LEADERSHIP AND COMMAND VIS-A-VIS ENEMY ACTIVITY; PREPARING THE ENTIRE NATION TO RESIST AND COUNTER FRENCH COLONIALISM

After the August revolution succeeded in giving birth to a people's democratic nation, under circumstances in which research, ingenuity, and expansion of the cryptographic net competed with providing for the training and development of cadre and personnel and solidly building a system of organization, the army cryptographic organization had to ensure the transmission of the content of leadership, direction, and command of the various echelons through the communication media, responsive to the urgent mission situation of our homeland. The cryptographic organizations from the 16th parallel [Dividing line set for British and Chinese troops entering to disarm the Japanese forces at the end of World War II. Tr./Ed.] up concentrated on serving leadership and command coping with the tricks and schemes and the actions to oppose and destroy or overturn the revolutionary authority on the part of the Chiang bunch and domestic reactionary gangs. The cryptographic organizations in Hanoi, Phuc Yen, Vinh Yen, and Phu Tho ensured the transmission of secret messages directing the struggle, and suppressing subversive acts by the Nguyen Hai Than, Nguyen Tuong Tam, and Vu Hong Khanh gangs; secret message directed provisional political arrangements between the Viet Minh front and the Viet Cach and Viet Quoc.

On 23 September 1945, when the French colonialists, aided by the British, opened fire and occupied Nam Bo, striking and taking over communication centers, the organizations and armed units of Nam Bo overcame obstacles to protect the evacuation of radio stations into bases, ensuring liaison within the region [vung] and with Central.

At 0815 on 25 September 1945, Nam Bo Cryptographic enciphered a message reporting to Central Party and the government concerning the resolution of the Nam Bo Sector Committee [xu uy], determined to resist the French colonialists. At 1010 the same day, the comrades received and deciphered the Central Party instruction agreeing with the resolution of the Sector Committee and the Nam Bo Resistance Committee. The cryptographic units in the South ensured the transmittal of the secret messages of the High Command, instructing the Southern military groups to advance and serve in the resistance struggle against the French in the South.

The task of enciphering and deciphering messages at this period was still elementary, but a first step toward an orderly routine. A number of regulations on enciphering and deciphering were issued, aimed at protecting technical secrets and the content of secret messages. All cryptographic materials had to be reduced to bare bones and constantly kept at the side of the cryptographer. The element enciphering and deciphering messages was to be compartmented by net, and not to express curiosity to know the contents of the work of another, between elements, or between each individual in his special compartment. Ordinarily, messages were transmitted by radio or post. In the beginning, for messages sent by post, the procedure was that the place sending and the place receiving were written in the clear, with the content alternating, some sections clear, some in cipher. If the messages was going by radio, then the sending place, receiving place, and precedence were written in French. Upon reexamination, these arrangements were seen not to be of value in protecting secrecy, thus gradually redone. Because going and coming was difficult, the replacement of the types of systems in use was not effected at an exact time, nor were new systems distributed in timely fashion.

The situation became more urgent with every passing day. On 19 October 1946 our Party's army-wide military affairs conference resolved clearly that "We must conclude absolutely that, sooner or later, the French will strike us and we absolutely must strike the French." As a result, at the General Staff and the units, the volume of secret messages to encipher and decipher increased daily, with higher precedences.

On 20 November 1946, the French colonialists opened fire, attacking and occupying Hai Phong and Lang Son, increasing the landing of troops at Da Nang, and staging many provocations in Hanoi. The army cryptographic organizations ensured timely encipherment and decipherment of the leadership guidance and command content from Central and the High Command to the theaters, and enciphered and deciphered the situation reports of the theaters to Central. Thus the operational experiences at Hai Phong, Lang Son, Nam Bo,etc., were quickly sent to other regions to study and apply.

On 16 December 1946, the General Staff cryptographic organization was tasked to encipher a message from the Standing Committee of Central Party to the provincial party HQ in the South, with contents as follows: "According to the situation on the French side and the greediness of the colonialists, there is only one global war, protracted, sharp, with difficulties newly resolved for the sovereignty of Viet Nam. The Party guideline is that it is absolutely essential to prepare. We must have good cadre and masses. Fully understand protracted resistance. Somehow victory will come to us."

In accordance with instructions from Cde Hoang Van Thai, the General Staff cryptographic organization reexamined the cryptographic system arrangements in the units and aligned and allocated cadre and personnel prepared and on duty to perform the mission, so as to quickly encipher or decipher instructions and operational orders of the General Staff, going to the combat sectors and the fronts.

Before the French military provocations in Hanoi took place, the General Staff cryptographic organization, along with the General Staff organizations, moved from 26 Hang Bai Street out to Thai Ha Ap (some 200 meters southwest of Dong Da Hill).

On the night of 18 December 1946, the French military command sent a letter of ultimatum to our government, calling for the stripping of weapons of self-defense and the occupation of the Hanoi Office of Public Security.

Implementing a resolution of the Central Party Standing Committee, the High Command [Bo Tong chi huy] issued the order to open fire and uniformly attack the French military on the night of 19 December 1946. From dawn on the 19th, Cdes Hoang Van Dong and Luong Dzan received orders to carry knapsacks and cryptographic systems up to a special place in the sector and work for the chief of the General Staff (at Thai Ha Ap) in order to await orders. And Cde Hoang Van Thai instructed: All cryptographic cadre and personnel who did not have to go to a sector to work were to stand by and be prepared to receive assignment.

At 0800 on the morning of 19 December, Cde Hoang Van Dong and Cde Luong Dzan received the task of enciphering an immediate message (received with a note and signature of Cde Hoang Van Thai, "need to encipher at once") going to the units, text as follows:

The French aggressors have issued an ultimatum disarming our army, self-defense, and public security. Our government has rejected this ultimatum. Therefore, at the end of 24 hours the French aggressors will definitely open fire. Instructions from Central: All will be prepared!

On the heels of which, this order from the Minister of National Defense/Commander-in-Chief to the entire military:

The motherland is endangered! The hour of combat has arrived!

Per instructions from President Ho and the government, and as Minister of National Defense and Commander in Chief, I order the entirety of the Vietnamese national army [bo doi Ve quoc quan] and self-defense militia [dan quan tu ve], Central-South-North, to the man, to rise up.

You must rush to the front, kill the aggressor, save the nation.

Give your life in battle, to the last drop of blood!

Exterminate the French colonialist gang.

Be resolved to fight!

Vo Nguyen Giap

The two message texts above were speedily encrypted and sent to the combat sectors and fronts.

Immediately thereafter, flash [hoa toc] messages were enciphered and sent to combat sectors 1, 2, 3, 4, 11 and to Da Nang, with the request that they be in the hands of the command comrades of the sectors and fronts prior to 0930 on 19 December 1946 in order to implement the order to open fire and strike the enemy on a coordinated basis on the theaters of war at the precise hour determined, contents as follows:

"The freight will arrive at 1800 hours 21 December 1946. The freight carries the code symbol A + 2 and B-2. Pay attention and meet the freight at the exact time."

Code symbols A + 2 and B-2 were previously established by the General Staff with the sectors. A was the hour, B was the day of the attack. A + 2 was [the stated time,] 1800, plus 2, or 2000 hours. B-2 was [the stated day,] 21 December minus 2, or the 19th.

At 2000 hours on 19 December 1946, at the High Command [Bo Tong chi huy] organization, which had relocated at that time to Chuong My (Ha Dong), Cde Commander-in-Chief [Tong chi huy] Vo Nguyen Giap stood on Chua Tram mountain along the Mai Linh river looking toward Hanoi and waiting for the signal of our guns. The cadre comrades and soldiers of the operations, communication, and cryptographic organizations also burned with impatience, waiting for the results of their service in transmitting command orders to open fire and strike the enemy.

At exactly 2003 on 19 December 1946, electric lights in the sky over Hanoi suddenly went out. Salvos resounded from the fortresses at Lang, Xuan Canh, Xuan Tao, etc., raining down on the heads of the French aggressors in the strongpoints they had set up in the city.

The Cde Commander-in-Chief was quite satisfied with the results of sending the combat orders of Central and the High Command to the units, guaranteeing timeliness, secrecy, and precision in terms of command organization and technical means which were limited.

Keeping pace with the militia of the Hanoi capital, the army and citizens in the large cities and regions, such as Nam Dinh, Vinh, Hue, Da Nang, Bac Giang, Bac Ninh, Hai Dzuong, etc., also opened fire, striking the French aggressor forces. The military cryptographic organizations in these places ensured accurate, secret, and timely encipherment and decipherment of orders and combat instructions from upper echelons.

Thus having taken the step to war, the army cryptographic organizations, from cryptographic organizations of the combat sectors, fronts, and especially the cryptographic organizations of the Capital, Thang Long, Son Tay, Ha Dong, etc., regiments requirements, did a good job of accomplishing their task of ensuring the service of leadership and command in striking the French aggressors, although the units lacked people and technical means. Via the system of cryptography and communications, the Ministry of National Defense, the High Command [Tong tu lenh], and the General Staff were able to grasp the situation in the theaters of war, and leadership, direction, and command from the Central Party and the army were timely, vis-a-vis the regions [dia phuong] and units throughout the entirety of the nation.

Having come through fifteen months of building organization and technique, while having to serve in ensuring the secrecy of the content of leadership and command from the Party and army via the means of communication, although faced by very many difficulties in this initial period, the army cryptographic branch strove upward to do a good job of accomplishing their mission, handed to them by the Party and army. The principal determining factor was the concern shown by the upper echelon leadership cadre, always creating [favorable] conditions and wholeheartedly assisting the army cryptographic organization in building organization, technique, and its job of serving leadership and command. Cdes Pham Van Dong, Vo Nguyen Giap, and Hoang Van Thai, etc., in the first stage of establishing the army cryptographic branch, not only provided concrete guidance concerning the direction of the job, but also personally introduced and assisted in the selection of trustworthy people--personally commented, gave suggestions, sought out documents for the cryptographic organization. After the July-August 1946 Fontainebleau Conference, Cde Pham Van Dong returned, bringing a French book on cryptography, *Le Chiffrement et le dechiffrement*, by Jean Bubois, and passed it to the cryptographic organization for research and reference. Although these accomplishments were only the first step, they demonstrated patriotism, self-reliant will to create technique, and minds determined to accomplish the mission, on the part of the initial contingent of army cryptographic cadre and personnel.

These results and accomplishments have extreme significance, in that they helped the branch extract lessons from real life experience and participate in the building and combat of the branch.

Notes

1. Based on the 7 September 1945 directive concerning the establishment of the General Staff, Ministry of National Defense (in the History of the General Staff).

2. Now 18 Nguyen Dzu Street, Hanoi

3. Cde Ta Quang De (Ta Quang Dam) is a patriot intellectual who had been a district chief, a scoutmaster in the old Boy Scout movement, and in the student and youth movements before the August revolution. Cde De was given the responsibility of chief secretary of the Communications-Liaison Bureau, especially in charge of cryptography.

4. Thuyen later took the nom de guerre of Hoang Thanh; he was a senior scout in Thanh Hoa, in the troop of which De was scout master. Cde Quan, alias Cde Hoang Van Dong, was chief of the Cryptographic Bureau from the end of 1946.

5. Cde Tung Anh intended to study and become expert, then return to the regional level, but he was retained afterward at the Cryptographic Bureau.

6. Original title, *Elements Cryptographic* (by Capitaine Baudouin). [Reference is evidently to the first, 1939, edition rather than the later, 1946, edition, which reflects the author's rank as major. A copy of this work, inscribed by the author to American cryptologist William F. Friedman, recalling a shared experience in World War I with the U.S. 32nd Infantry Division, is in the Friedman Collection, Marshall Library, Lexington, Virginia. -- Tr./Ed.]

7. The scouting movement of Vietnamese and Indochinese youths and students before the August Revolution involved games such as "Morse Code" and "Maneuver" (a big game). In the "Maneuver" game, the player had to solve secret letters written under the form of a simple cipher.

8. Each letter [piece of correspondence] had a group of digits used to make the cipher key.

9. Subordinate to the relief army newly organized when the French forces entered the North "to guard Japanese POWs," replacing Chiang's forces, who withdrew back home.

Chapter Two

Consolidating and Building Organization and Professional Technique, Meeting Command Leadership Requirements in the First Five Months of the Protracted Resistance (1947–1950)

THE ARMY CRYPTOGRAPHIC BRANCH BUILDS AND SERVES IN COMBAT IN 1947

As 1947 began, war had spread to many places. From the very first years of the resistance against the French colonialists, the MND-High Command and General Staff were settled on expansion of the armed forces and set the course for the activities and tasks of the army cryptographic branch. The resolution of the first Nationwide Conference on Military Affairs, meeting from 12-16 January 1947, put it clearly: ". . . we must pay attention to the immediate training of many cryptographic personnel. Don't constantly shift cryptographic personnel around." In a resolution of the Conference of Sector Chiefs, meeting in Viet Bac in March 1947, under "troop problems," and the objective of "organizing the specialty branches of the troops," is also stated: ". . . we must train personnel and organize communication among the troops by cryptography."

Implementing the resolution of the above conferences and directions of the High Command and the Chief of the General Staff, the army cryptographic branch carried out many means of expanding and correcting the organization and training of cadre and personnel, researching the production of cryptographic systems, and organizing their use in meeting the requirement for command secrecy by cryptographic technique when employing communication media.

In February 1947, with agreement of the High Command, the Cryptographic Bureau organized the third Army-wide Cryptographic Conference, the purpose being to strengthen and ensure liaison on all lines, between Central and the regions, and between the regions and units, with each other. The conference requested the Cryptographic Bureau compile basic theory documents on cryptography to disseminate for cryptographic organizations army-wide.

Implementing the conference's resolution and satisfying the work of "organizing communication by cryptography," in February 1947 the High Command Cryptographic Bureau opened the second Cryptographic Personnel Training Course. Named "Viet Bac," this class opened in the village of Yen Thong, Dinh Hoa district, Thai Nguyen. Room and board for the students and class sessions alike were in the homes of compatriots in the area. The syllabus for this class was more or less that of the first class ("Hoang Dzieu"), with special attention paid to setting aside more practice periods. Of the more than twenty students in this class, there were a number of female personnel, such as young Miss Kim Chi, Hoang Bao Khanh, Hoang Lan, Nguyen Thi Lien, Le Anh Phuong, etc.

When it moved up to the Viet Bac revolutionary base (early 1947) the Cryptographic Bureau settled, as a matter of urgency, on a place to set up shop, and afterwards continually carried out all aspects of the task of building the branch, requesting upper echelons to supplement with a number of cadre and personnel, and implementing production of the types of cryptographic systems to supply to the units.

As of June 1947, the Cryptographic Bureau set up a party cell comprising comrades Ho Ton Vinh, Hoang Van Dong, Luong Dzan, Vu To, Nguyen Chanh Can, and Hoang Manh Tuan.

On 19 July 1947, President Ho Chi Minh signed a decree concerning the organization of the top-level command organizations of our army, consisting of the High Command and the Ministry of National Defense. According to this organization, the Central cryptographic organization became two cryptographic bureaus: the Ministry of National Defense Cryptographic Bureau, with responsibility for encrypting and decrypting messages serving command and the rear services mission, at the same time researching cryptography and cryptanalysis. The bureau was the charge of comrade Dinh Loan Thuyen. The Cryptographic Bureau of the High Command had responsibility for encrypting and decrypting secret messages of command guidance for army operations, building forces and various other aspects of the task, while at the same time researching and producing cryptographic systems (supplied to units army-wide), training cryptographic cadre and personnel, and research concerning cryptanalysis--Cde Hoang Van Dong was bureau chief. The bureau consisted of these sections: the Research and Training Section, the Clerical Section, and the Encrypting-Decrypting Section. The troop strength of the bureau of this time was sixteen people. Formerly, getting people to come do cryptographic work was usually from the troops – afterward, all of the people were chosen from outside. Because one could not come right out with the conditions and requirements of the task, when they finally saw the strict demands of the task, a number of personnel wanted to get out of enciphering and deciphering, or asked for a change in assignment. Being short of cadre because of supplementing the units, the bureau selected a number of female cadre: the youngsters were industrious in their work, but, when the time came that there was a need to organize cryptographic elements to go and serve on the front, there were difficulties because of the lack of male cadre.

The Encrypting-Decrypting Section only had two teams, one in charge of encrypting and decrypting with the sectors in the South, one with the sectors in the North. The technique of encrypting and decrypting had advanced greatly, compared with the previous year. In the course of a year, the cadre and personnel had become rather well acquainted with the various types of cryptographic materials, and enciphering and deciphering was fast and went without a hitch. By April 1947, outside of the regular liaison points [dau moi], cryptographic liaison nets were opened and expanded with Sector 6, the Southern Resistance Committee, and several sectors in Nam Bo. From the middle of 1947 on, with hostilities spreading, the cryptographic organizations expanded and liaison by radio was much increased. The cryptographic organizations took care of encryption and decryption to serve leadership and command with a noticeable volume of traffic: Figured from the beginning of the year until September 1947, the total number of secret messages outgoing and incoming at the Cryptographic Bureau approached 2,700 official messages. Especially, during this period, liaison with units in the South was relatively ensured regularly and steadily.

The cryptographic correspondence task became more routine in all of the work of copying messages,sending, receiving, and holding secret messages, incoming and outgoing.

In the sectors, the cryptographic sections were strengthened another notch. The Cryptographic Section's principal mission was the encryption and decryption of secret messages. Besides this, a number of places had entrusted to them the making of cryptographic systems and the training of cryptographic cadre and personnel from their own unit.

From February 1947, the Sector 1 Cryptographic Section implemented consolidation and building of cryptographic organization in the units below regimental echelon: the first stage, consolidating and building cryptographic organizations of battalions subordinate to regiments; the next stage, from the Second Sector Cryptographic Conference (August 1947) until the unification that produced Inter-sector 1 Cryptography, at which stage the cryptographic organizations were extended to a number of companies. When the Sector main force battalions had cryptographic organizations, the Cryptographic Section arranged direct means of liaison with these battalions, but, with the task of researching and producing systems during this period, the Sector Cryptographic Section could then only take care of the regiments, essentially by estimating cryptographic system usage, rather than showing new initiatives, leaving the regiments themselves to see to the cryptographic systems for battalion and company echelons.

The cryptographic organizations in Sector 3 formed a vertical structure in their speciality although the liaison organizations were not expanded and under firm control. The Sector Cryptographic Section organized a cryptographic net down to battalion and a number of essential independent companies. Besides these there were still a number of points in cryptographic liaison, including joint units from the Route 5 Front, provincial Resistance Committees, and local activities of the intelligence and munitions [quan gioi] branches, that needed to organize liaison by military cryptography. The cryptographic teams of the 34th and 42nd regiments regularly kept in liaison with the Sector Command [Bo chi huy Khu] and liaison with a number of battalions (via the means of Communication-Liaison's telephones). Average daily volume of message traffic by cryptography, sector-wide, was 100 official messages. During this time, because of few people, the section chief normally had to assume responsibility for the clerical mission, training, and system research; they almost never had systems in reserve--when it became necessary to organize joint liaison with regiments outside the area of responsibility of the sector, then there were no systems available to make prompt arrangements (e.g., with the Vinh Phuc and Dong Trieu regiments).

Also in 1947, in Sector 2, although the cryptographic organization had a structure of vertical organization, there was only liaison with one another in the areas of encrypting and decrypting secret messages--there was virtually no exchange of experience, specialty inspection, or the like. At Sector HQ, the Cryptographic Section only had three to four people, because the specialized elements had not yet been set up.

Sectors 2 and 3 opened a class for training in cryptography: the bulk of the content concentrated on practice, but a portion of the remainder followed the syllabus of the HQ training course. Selecting people to go work in cryptography was rather deliberate and careful, and selection was from among educated military personnel, endorsed by the immediate command echelons. Each military person going to work in cryptography had to have sufficient papers, such as personal history, endorsement from immediate commander, obligation to serve and maintain the secrecy of cryptography for the period of two years at the minimum, etc., sent to the Intersector Cryptographic Section.

In 1947, in Sector 4, there was a unified cryptographic organization, sector-wide, but, from the standpoint of horizontal and vertical control, it was still immature--the cryptographic organization had yet to have basic units, especially cryptographic organization in the area temporarily occupied by the enemy. The sector cryptographic organization at this time comprised only three comrades, one of whom was in the HQ Hoang Dzieu class, sent back as an augmentee, but still inexperienced, so the consolidation of the cryptographic organizational structure in Sector 4 encountered many difficulties and expanded slowly, compared to other sectors.

In mid-year 1947, although a structure had been formed, the Sector 5 cryptographic organization was not yet under tight control. The comrades in charge of regimental cryptography were routinely sent off on other tasks. As a result, cryptographic security was not thoroughly maintained. Many of the fellows doing message clerical work did not yet have their act together on outgoing and incoming messages. As for cryptographic systems, they followed, for the most part, the Cryptographic Bureau models with little modification, such as using the Julius Caesar substitution method, afterward adding the Stook [sic] system, which was somewhat more advanced.

In March 1947, Sector 5 opened its initial cryptographic course, with about twenty students. Although still lacking in training experience and teaching material, nevertheless the requirement was met to extend the cryptographic organization in the theater of southern and central Trung Bo [Annam]. In order to correct the organization and establish the important parts of the task according to the instructions of the Sector 5 Command, in July 1947 the Cryptographic Section convened a sector-wide cryptographic conference. Problems dealt with at the conference were: correcting the system of organization sector-wide, defining the posting of cryptographic cadre and cryptographic personnel, lessons learned from the task of encrypting and decrypting, etc., doing their bit to help settle the organization and realize the mission of the Sector 5 cryptographic organizations. Thus by the end of 1947, cryptographic liaison between the Sector HQ and the units was speedier and smoother than before.

In Sector 6, extreme south Trung Bo, three quarters of the area was temporarily occupied by the enemy, and communication was both very difficult and dangerous. The cryptographic organization was lacking in cadre-in-charge--most of them had additional duties. The cryptographic liaison net from Sector down to five regiments (the 80th Regt., Phu Yen; 83rd, Khanh Hoa; 81st, Ninh Thuan; 82nd Binh Thuan; and 79th, Dac Lac) was firmly in hand, but there were still no few difficulties. Seeing that the Sector 6 cryptographic organization had many weak aspects that had to be shored up, in accordance with a May 1947 proposal of the High Command Cryptographic Bureau, HQ decided to appoint Cde Nguyen Van Dzanh (a Cryptographic Bureau Cadre augmenting Sector 5) to go down and take over the Sector 6 Cryptographic Section.

Also from May 1947 the Cryptographic Section was separated from the secretariat and made directly subordinate to the Sector Command [Bo chi huy Khu]. In June 1947, the Sector 6 Cryptographic Section organized a sector-wide cryptographic conference to discuss the matters of improving the organization, replacing cryptographic systems, unified operating procedures, and the organization of a direct cryptographic liaison net with HQ. In August 1947 the Sector Cryptographic Section opened a cryptographic training class, adding to the sector headquarters organizations and a number of independent regiments in the area. With the regiments in the area temporarily occupied by the enemy, the Cryptographic Section drew a number of cryptographic materials from the Hoang Dzieu course to send to the fellows performing cryptography to research and study themselves.

Vis-a-vis the Nam Bo theater--a theater with "no forward area, no rear area"--with dense jungle, honeycombed with canals and streams, and enemy blocking posts, the need for cryptographic organizations had to overcome many obstacles in solving the problems of disseminating cryptographic systems and training cadre and personnel to respond to mission requirements. Conditions did not permit the Nam Bo cryptographic organizations to come up and attend the cryptographic training classes and cryptographic conferences at Central, so the exchange and acceptance of general experiences of the branch were limited. To overcome this limitation, the Cryptographic Bureau issued guidance and professional exchanges by message, and took advantage of leadership and command comrades being posted to Nam Bo to send along a number of models of cryptographic systems. Under the guidance of HQ, Nam Bo, the military cryptographic branch in Nam Bo progressed step by step.

Concurrent with the tasks of building organization, training cadre and personnel, and arranging liaison nets, the army cryptographic branch advanced in research and promulgation of cryptographic technique. Systems were all improved with respect to content and format. Chart systems were constructed more scientifically. The plain content of the chart was enriched, the display clarified and made convenient for encrypting and decrypting. The cipher strip had many columns, the movable horizontal strips had many scrambled rows of cipher letters in order to produce polyalphabetic substitution, speed up things, and increase security. The seven-column system was enhanced in quantity by the method of using irregularly and in rotation the ten columns in twenty-six random, unrepeated, alphabets, thus changing the appearance of the cipher text to a rather high degree. This type of system was arranged and brought into play in the liaison net between HQ and the sectors, with the designation "Glori-CK."

Initial specialist-task discipline was built with a number of regulations: work "close by" the person in command; when you are working, you must be secretive; you must keep the content of secret messages just as securely as you do your cryptographic systems, etc.

These initial advances in organization, professional technique and routinizing the task demonstrated that the army cryptographic branch was filled with the spirit of the instruction from the Standing Committee of the Central Party: "Frequently change cryptography and the hours of contact of the radio stations; militarize the radio station and cryptographic organizations."[1] This was also the basis for the success of the branch in realizing its mission of maintaining command security by cryptography when we moved into combat to defeat the assault by the French aggressors into the Viet Bac revolutionary base in the fall and winter of 1947.

In October 1947, the French colonialists concentrated a large force, consisting of 20,000 men, forty aircraft,and eighty motorized vehicles, to open a three-directional assault on Viet Bac, aiming to "strike our nerve centers of resistance, wipe out our main force units, [and] have conditions in which they could install puppet authority on the spot."[2]

On 15 October 1947, the Standing Committee of the Central Party issued instructions to demolish the winter offensive by the French aggressors, and, at the same time, proposed the mission and delineated the course of action by the military and people of Viet Bac and our entire nation.

On 27 October, the Cryptographic Bureau cadre and personnel of the High Command conveyed Order No. 132 from the High Command to the sectors and units nationwide: "Strike hard on the Song Lo and Route 4 front; destroy transportation supplying the enemy, set up ambushes on the jungle roads, strike the river routes;constantly harass the enemy bases, encircle and eliminate small positions. Sectors are to strike hard in order to coordinate with Viet Bac."

On its heels was a secret message from the Cde Chief of the General Staff assigning responsibilities to the units as follows:

> "On the Rte 4 front, the 74th Regiment (Sector 1) is in charge of the Cao Bang front, with the Lang Son Regiment (Sector 12) having the mission of striking hard on Route 4 from Lang Son to Dong Khue. The 2nd Battalion of the 350th Regiment of HQ has the mission of striking the enemy on the Lang Son - Vo Nhai Road. The 232rd Battalion of HQ blocks Binh Gia Street. The 80th Battalion of HQ is stationed at Trang Xa, Dinh Ca.

"On the Song Lo front: the 112th Regt (Tuyen Quang), the Sector 10 artillery units, and the 18th Bn of HQ have the mission of striking the enemy transportation on the river and on foot.

"The 72nd Regt and the 19th Bn (Sector 1), the 102nd, and the 160th Bns of HQ have the mission of striking the enemy on the Route 3 front.

"On the Tuyen [Quang]-Thai [Nguyen)] road front, use the 350th Regt of HQ, two regional battalions, and the 103 Bn of HQ (if necessary, these can be augmented with additional mobile forces) . . . ."

The cryptographic organizations throughout the military labored unmindful of day or night, transmitting as a matter of urgency the instructions, orders, and reports, secretly, swiftly, and accurately.

During this same time, implementing instructions from the Chief of the General Staff, the Cryptographic Bureau divided its forces according to the organizations of the High Command and General Staff and prepared to arrange people and means to perform the mission in various directions.

The force to remain behind in Dinh Hoa (besides having to ensure liaison with the light element of the High Command) was responsible for the cryptographic liaison net with the units subordinate to southern Trung Bo and Nam Bo.

The force to accompany the important greater portion of the HQ organization moving up to the new site (Don market, in Bac Can) was responsible for liaison with units in Bac Bo [Tonkin] and Sector 4. At the same time, the Bureau assigned three cryptographic teams to go and serve in three places: one team, under Cde Luong Dzan, to accompany Cde Commander-in-Chief Vo Nguyen Giap personally studying the Route 4 front in order to derive general experiences to direct the other fronts; one team to go with Cde Chief of the General Staff Hoang Van Thai down to direct the preparation of a secure base sector for the Central Party and the government, to block the enemy pushing down toward Chu market, directing the evacuation of workshops and depots around the town of Bac Can, this team being under Cde Hoang Manh Tuan. The third team, under Cde Tran Dien, went to serve alongside the Central Party organizations.

Having to organize many task elements when the cadre and personnel were few, having to organize cryptographic liaison nets over a wide area and to coordinate many units, with cryptographic message volume more than before, still the cryptographic cadre and personnel sensed the significance of the battle, made many efforts that surpassed expectations, realized the encryption and decryption to ensure that the transmission of each order, directive, and report was accurate and timely. In the sectors and fronts combined, the cryptographic organizations also performed their mission well. The army cryptographic branch diligently and efficiently served the leadership and command comrades at all levels, participating in the glorious victory that shattered the French aggressors' attack, guarded the security of the Viet Bac base, kept intact and expanded [our] forces, and held the resistance base of the whole nation.

The army cryptographic branch passed the test and grew another notch, deriving many useful experiences in the mission of combat service, on a far-flung front, responding to the requirement of our army to speed up mobile warfare.

EXPANDING ORGANIZATION – STEPPING UP THE TRAINING OF CADRE AND PERSONNEL

The Fall-Winter victory of 1947 in Viet Bac ushered the resistance of our people into a new stage.

So as to consolidate direction from the Central party and Main Military Committee in the new stage, the Central party determined to strengthen the directing organizations at the various levels, especially the military organizations. On 25 January 1948, the President of the Democratic Republic of Viet-Nam issued a decree organizing and unifying the sectors, to create "intersectors" from north to south: the seven sectors in Bac Bo [Tonkin] merged into three intersectors, intersectors 1, 10, and 3. The four sectors in Trung Bo [Annam] made two intersectors, intersectors 4 and 5. In Nam Bo [Cochin-China], there would be three sectors, 7, 8, and 9, and the Saigon-Cholon Special sector. One element of the main force military was broken up to produce independent companies, armed propaganda units, and volunteer sections for deep penetration into areas occupied by the enemy, to mobilize guerrilla warfare. Consolidated battalions were organized with the mission of mobile operations to sap enemy strength, and create conditions for the spread of guerrilla warfare. The cryptographic task was also realigned, consistent with the organization and command requirements of the High Command, the Intersectors, and the units.

The Intersector Cryptographic Sections were established on the basis of unifying the Sector Cryptographic Sections, while, at the same time, cryptographic organization in lower- level units was also being strengthened, in accordance with the spirit of the 30 January 1948 General Staff instruction: "Organizations must militarize, retaining only the essential elements . . . the troops must be lean and orderly, suitably equipped . . .relying on cadre ability and standards, and our equipment capabilities . . . ."

In order to participate quickly in strengthening the organization of cryptographic activities of HQ and the Intersectors and units, and to meet the requirements of the theaters of war, in April 1948 the Cryptographic Bureau itself opened the third class for training cryptographers. This class took the name the Bong Lau class. Sessions were conducted in the village of Yen Thong, Dinh Hoa district, Thai Nguyen. From northern Trung Bo up, and from units in the High Command, forty-five students were selected to come study--the majority of these being comrades who, although they had worked in cryptography, had never received school training. Cde Hoang Van Dong was in charge, and a number of Cryptographic Bureau cadre were appointed instructors. The classroom was rather tightly organized. The syllabus was improved and at a higher level than previous classes. Besides the part on training in the professional speciality, special emphasis was placed on training in grasping the responsibilities of the mission, and making routine out of the task of organization and cryptographic cadre and personnel at the various levels. Students researched and practiced making cryptographic systems, researched the frequency of the Vietnamese language, cryptanalytic methods, and principles of security. In the final examination, besides questions about classic cryptographic theory and cryptanalytic practices, each student had to personally produce a code chart [bang luat]. After two months of study, it was essential that the students from units and intersectors return to those units and intersectors. Upon return to their units, the students spread the results of their training. Many comrades were selected to take charge of regimental cryptography or as team chiefs for research and training in the intersector cryptographic sections.

Of this class, three comrades, Nguyen Van Dzuyet, Tran Cong Ta, and Le Nhan, became augmentees for the cryptographic organizations of the Party-Government system.[3]

After their establishment, and, as a matter of immediate attention, the Cryptographic Sections of the intersectors commenced to build organization responsive to mission requirements in accordance with their function.

The Intersector 1 Cryptographic Section came into being in February 1948, comprising the merged Sector 1 and old Sector 12 Cryptographic Sections, quickly settling the organization and consolidation of the Intersector's subordinate cryptographic organizations. Cde Ngo Vi Thien was Section Chief. Most important of the Intersector's cryptographic tasks in 1948 was the organization of cryptography for units subordinate to old Sector 12 and Intersector HQ. The Intersector Cryptographic Section took its structure from elements on the spot, such as those of the former Sector 1. As task requirements and the number of personnel increased, the enciphering-deciphering element was divided into two teams, secretariat and enciphering-deciphering. The dividing of specialized responsibilities in the Section was made clearer, the work of the two teams went deeper into specialization, orderly routine, increased productivity in the task.

The Intersector Cryptographic Section organized many conferences to solve problems of organization, technique, and professional knowledge vis-a-vis cryptography intersector-wide; organized and mobilized emulation movements, raising to a feverish pitch the working atmosphere from Section down to the organization of unit cryptographic. The plan to rectify the organization and expand cryptography to the units was constructed in parallel, with concrete provisions, divided by time periods for realization. At the beginning they concentrated on rectifying the regimental cryptographic organization. Afterward, cryptographic organization was expanded to the battalions and component companies of the regiments. This work met no few difficulties, for cryptographic personnel were insufficient, and a number of unit command comrades had yet to perceive the necessity of expanding cryptographic organization.

With the direction and support of the Intersector Cryptographic Section, the cryptographic organization of the regiments was built single-mindedly, by seeking out ways of realizing the plan. In order to have a sufficient quantity, and to raise the level of cryptographic cadre and personnel for directly subordinate units, all regiments opened training classes in the use of cryptographic technique, averaging ten to fifteen days each.

From February 1948 to January 1949, the regiments opened eight classes, with nearly sixty students.

The Intersector Cryptographic Section also launched an effect to foster cryptographic cadre-in-charge for the units, comprising ten comrades studying for one month. In nearly two years, Intersector Cryptography had held twelve classes for nearly 100 cryptographic students.

Notwithstanding, the mission of developing cryptographic cadre and personnel in Intersector 1 still floundered and was hindered by difficulties: the criteria for selecting people to go work in cryptography initially were boiled down to just one consideration--a cultural level equivalent to the first class, vouched for by the unit command section, but realistically the majority only had a level of elementary education, or second class. And through practice of the mission, Intersector 1 Cryptography still showed weak aspects of hit-or-miss training conditions and no unified program--not yet organized to summarize and publicize specialist lessons learned – not a few of "the cryptographic personnel weren't keen on the cryptographic mission because they'd been pushed into it and saw progress as slow, principally in battalion and company cryptographic."[4] Between an atmosphere of combat seething throughout the army and a breast full of enthusiasm on the part of young people, there were those whose aspiration was to bear arms to kill the aggressor.

The administration of cryptographic cadre and personnel was also initially determined: appointing a regimental-level cryppie as recommended by the Intersector Cryptographic Section and decided upon by the intersector HQ; appointing cryptographic team chiefs under routine orders of Intersector decision. But at regimental level, wishing to propose appointment of battalion and company cryptographic personnel remained extremely difficult, because the majority of the mates had concurrent responsibilities, holding different assignments (the bulk of the cryppies were also company clerks).

Enciphering and deciphering continued as the essential task – serving combat command by means of cryptographic technique in battle. The cryppies accumulated on-the-job experience, avoiding many mistakes in enciphering and deciphering, emulating precise encipherment and decipherment, and seeking out methods of decipherment of messages containing many groups enciphered in error. The emulation drive between units, monitored by the Cryptographic Section, was mobilized beginning August 1948, and pursued monthly, with observations and results communicated to all units, etc., resulting in visible progress in specialization capability on the part of cadre and personnel, generally speaking.

In January 1949, the third Intersector-wide cryptographic conference concluded that the most important task was to strengthen every aspect of Intersector cryptography, with special attention being paid to developing the branch's technique. Responding favorably to the emulation drive mobilized by the branch to produce innovation and improvement in technique, at Intersector [level] a research team was created--one of three essential teams in the Cryptographic Section.

[After] more than three years of building and strengthening the organizational system from Intersector Cryptographic Section down to the unit cryptographic organizations, the Intersector 1 cryptographic organization (and before that the Sector 1 cryptographic) observed themselves:

1. From a tiny organization, careless and casual, until now, the Intersector 1 cryptographic branch has developed into indispensable units and is advancing rapidly on the path of development in order to grow with the national army.

2. The specialist assignment has advanced a great deal, enciphering and deciphering rapidly, and with innovations, having new products, so as to be separated from the cryptographic systems previously in use.

3. [With] personnel from six to seven people, weak in capability, to the present expansion - 100 cryppies, the majority trained. Cryptographic cadre in the units also have adequate capability in the specialty task. As a result, the cryptographic branch has progressed and its capabilities will progress even more as a result of the strenuous emulation effort:

● Rectifying organization sensibly

● Thoroughly grasping and leading the units in the speciality task

● Researching and developing...[5]

In Intersector 3 in February 1948, the Cryptographic sections of Sectors 2 and 3 merged to create the Cryptographic Section of Intersector 3. The Intersector 3 Cryptographic section quickly established a system of organization and unification for the entire Intersector. Cryptographic organization of the intersector comprised: at Intersector HQ, the Cryptographic Section, under Comrade Le Hai; Cryptographic Sub-sections at regimental level; cryptographic teams at battalion; and cryptographic sub-teams [tieu to] at company level.

The Intersector Cryptographic Section was an independent section directly subordinate to the Intersector HQ. Troop strength was increased through the merger, building the foundation for organizing Intersector cryptography and quickly firming up the task.

After 1948, following their defeat in Viet Bac, the French aggressors turned their strength to the [Red River] Delta region in a policy of "squeeze and spread the oil slick" to protect this important theater of war.

[As to] serving command in a broad area of responsibility: the Delta is a place of dense population and much property – the enemy made a push to destroy the revolutionary bases and capture cadre and guerrillas, win the people, and plunder the property. The mission of the intersector cryptographic organization was to ensure command secrecy by cryptography via radio, the content being guidance for carrying out the frustration of the enemy's new trick and schemes, striking guerrillas, destroying the puppet administration, serving combat operations--quite a bit of message volume. Each day the Intersector HQ had around 300 official messages outgoing and incoming.[6] With a lot of messages, the principal task for the Section centered on encryption and decryption. The Cryptographic

Section had the mission of directly handling official messages with the bureaus and sections of HQ and also with external organizations (e.g., the Resistance Committees, etc.).

Right from the time of its establishment, Intersector 3 cryptographic become the concern of Intersector HQ, creating conditions from a material [standpoint] while mobilizing the intellect: HQ had instructions concerning the cryptographic task sent to unit command sections. The principal content of the instructions brought up the essentiality of liaison by means of cryptography; selection and appointment; policies toward people performing the cryptographic task, etc.

From mid-year 1948, in the 42nd, 64th, and 34th regiments, cryptographic organization had developed down to the companies, the provincial resistance administrative committees, and the armed propaganda units. The regiments had contact both by radio and wire, thus, between Intersector and regiments, close, solid liaison was assured. If, at the Intersector, the research task was constrained by a volume of cipher messages that had to be gotten out, the comrades in charge of cryptography down at regiment were able to participate constructively in research and to produce systems through their own activity.

The Intersector Cryptographic Section organized a talkfest, a form of specialist activity appropriate to the time, with the object of mutual assistance in the specialty task between the cryptographic organizations of the Intersector, creating circumstances for the unit cryptographic organizations to comprehend and bond tightly, uniting with one another in the assignment, exchanging and supplementing each other's experiences.

By 1949, as a result of changes by the Intersector HQ, the secretariat [van phong] of the HQ was turned into the Clerical Bureau [phong bi thu] of HQ, Intersector 3, the Cryptographic Section becoming one of two principal sections in the Clerical Bureau, managed by the chief clerk as far as administrative matters were concerned, and by the Intersector HQ as far as the specialty task was concerned. Troop strength of the Cryptographic Section was twelve people, as when the Intersector was formed, with a seven-man Encrypting-Decrypting Subsection divided into an encrypting-decrypting liaison element with HQ and an encrypting-decrypting element with the regiments, etc. The clerical subsection had three people copying, receiving, and sending messages and papers. It must be said that the cryptographic teams of the regiments subordinate to the Intersector were developing greatly during this period. Component battalions, companies, and special units engaged in armed propaganda in the enemy areas all could use cryptography. Besides ensuring and arranging cryptographic liaison nets in the military system of the intersector, the Resistance Administrative Committee, the armed propaganda units, and detached guerrillas, liaison nets also had to be open with intelligence, munitions,the provincial units, the Command sections of the fronts, etc. The Intersector, in particular, set up an additional liaison net with the Route 5 front to ensure firm, constant contact.)

The number of liaison points having thus grown, each cadre and person had to look into the work of encrypting and decrypting and constructing new cryptomaterials adequate to satisfy the immediate need, and that in the short term. Each time Cryptographic had to supply cryptographic materials for radio stations popping up from nowhere was a test of the Intersector cryptographic organization, having to surge to overcome obstacles and accomplish the new mission.

A routine horizontal check of the Intersector branch having taken shape, there had to be a concrete check of subordinate levels. The Intersector Cryptographic Section checked regimental cryptography; regimental cryptographic checked that of the battalions and companies.

In 1949 the Intersector Cryptographic Section held thirteen training classes, both in-service and classroom. Through these classes they created conditions for 75 percent of the total of 130 cryptographic cadre and warriors in the Intersector studying to raise their proficiency.

****

In September 1948, the Intersector 10 Cryptographic Section was set up, after Sector 14 merged with Sector 10. Comrade Ngan Ba Hong was assigned as section chief. The Intersector 10 Cryptographic Section convened an Intersector-wide cryptographic conference to unify and perfect the organization, reorganize the cadre, and replace cryptographic materials.

At the Intersector level, the Cryptographic Section formed three subsections:

- Encryption-decryption Subsection

- Clerical Subsection

- Training and Techniques Research Subsection

Cryptographic organization at regimental level was unified and called the Cryptographic Subsection; at battalion and company, it was the Cryptographic Clerk.

Responsibilities of the Cryptographic Section were determined at the Intersector Cryptographic Conference as ensuring encryption and decryption for the Intersector HQ; building a cryptographic liaison net system for the entire Intersector; supplying cryptographic materials and systems for the independent regiments and battalions; and holding classes for cryptographic cadre and personnel.

Around the end of 1948, the Intersector 10 Cryptographic Section opened a supplemental class for cadre in charge of cryptography in the Intersector, taking the name "Nguyen Van To" for the class, which comprised twenty student comrades. The 148th and Song Lo regiments also held classes to improve cryptographic personnel for battalion and company echelons.

Besides their principal responsibility for encrypting and decrypting, the regimental cryptographic subsections also had a regular responsibility doing a frequency count of the Vietnamese language, in accordance with a work plan and directions from the Intersector Cryptographic Section.

****

After 1948, the cryptographic organizations of Intersector 4 were consolidated and expanded.

At Intersector HQ there was the Cryptographic Section under the HQ secretariat. The Section Chief was Ho Si Bang. The Intersector Cryptographic Section had three teams: clerical, encrypting-decrypting, and research and training. The encrypting-decrypting team had two elements: the element encrypting and decrypting within the sector, and the element encrypting and decrypting external to the sector.

Regiments had cryptographic teams directly under the staff, Command Section, or political commissar. The cryptographic team had around three or four people.

At battalion-level cryptographic, there were two people, and one at company level. In 1948, Intersector cryptographic expanded to the battalions and a number of independent companies.

On the Hue-[Quang] Binh-[Quang] Tri-[Thua] Thien front, there was established a Binh-Tri-Thien Sub-Sector Cryptographic Section, with a liaison net comprising the 101st Regiment, 95th Regiment, with Intersector 4, Intersector 5, and the High Command [Bo Tong Chi Huy].

After the battle of Hoi Mit, because of the compromise of cryptographic techniques of organizations outside of the army, the Intersector [Party] Committee convened an Intersector-wide cryptographic conference to unify and standardize the cryptographic task. A common cryptographic organization was established for the entire Intersector (embracing the cryptographic responsibility of the army, the resistance, public security, intelligence, etc.). This organization had the responsibility to make cryptographic systems to use in the Intersector.

As for the cryptographic organizations in Intersector 5, the High Command augmented cadre to go down and help consolidate and strengthen in various aspects.[7]

Aiming to correct the shortfalls in communication liaison and see to daily improvement for command guidance at the various echelons, according to a proposal from the Sector 5 Cryptographic Section, and with the concurrence of the government representative in the South and that of Sector HQ, on 24 and 25 February 1948, a Southern Trung Bo [southern Annam] cryptographic conference (sponsored by Sector 5) was convened. Attending the conference specially was Cde Pham Van Dong, the government representative, and the gentlemen who headed the secretariat and radio of the South, the post-telecommunications director of the South, representatives of Sectors 5, 6, and 15 cryptographic, the Resistance Administrative Committee of southern Trung Bo, representative of the Southern Trung Bo Liaison Bureau, etc.

The conference discussed means of consolidating the system of liaison in Southern Trung Bo; the unification of the cryptographic branch in the sectors; alleviating some remaining shortcomings, such as the relationship between the postal and cryptographic branches in the matter of message precedence indicators, etc.

In accordance with the 25 January 1948 decree, as of September 1949 Intersector 5 was officially established, comprising Sectors 5, 6, and 15. At that time the situation with respect to development of the cryptographic organizations was not uniform. The Sector 5 cryptographic organization was relatively routinized and methodical, whereas Sector 6 was still undergoing consolidation and still divided into military cryptography, political cryptography, etc., and Sector 15's cryptographic organization was just taking shape, for Sector 15 was established last. Thus an orderly cryptographic task in Southern Trung Bo had not yet been built through unification, relating tasks between the cryptographic and postal organizations; radio was not yet tight in hand; the sending and receiving of messages was still error-prone and late. Confronted by this situation, the High Command appointed Cde Nguyen Chanh Can, a Cryptographic Bureau cadre, to go down and assist the Intersector cryptographic organizations in straightening itself out and in increasing liaison between the Intersector and HQ.

In Nam Bo, at the end of 1948, Cde Vu To, a cadre of the Cryptographic Bureau, was assigned by the High Command to go down to Nam Bo to "grasp and comprehend thoroughly the cryptographic situation in the sectors and to hand over cryptographic materials,"[8] and to strengthen the organization and the technical cryptographic skills of the units of the Nam Bo theater. To that end, he was appointed chief of the Nam Bo Cryptographic Section, "although a cadre from Central, still quite young."[9] After being strengthened by cadre sent down from the Cryptographic Bureau, the Nam Bo cryptographic organization systematically progressed in its building and expansion. The situation involving the cryptographic task was part and parcel of the general situation of the staff task in 1948, that being "a year in which the staff task had to pay much attention to organization to realize Central's program for expanding guerrilla resistance all over the place, not just in the main theater but beginning to pay attention to the theaters of [Quang] Binh-[Quang] Tri-[Thua] Thien and Sector 5, Nam Bo, Laos, Cambodia, etc. Thus this was a fresh task . . . not yet having satisfied the heavy [original] mission entrusted from above . . . The system of communication liaison with the sectors and provinces was not yet coordinated wide scale, principally with distant theaters."[10]

In order to overcome the difficulty of a situation in which a theater was split up, yet leadership and command had to be assured on a timely basis, in 1949 Sector 8 organized radio station equipment and organized cryptography in the various regiments and battalions, such as the 310th and 925th battalions of the 99thRegiment and in the military intelligence units.

By May 1949, the system of liaison by radio station and cryptography of Sector 9 with the High Command was fully realized.

In July 1949, liaison by cryptography was fully realized directly between Sector 7 and the High Command, rather than through an intermediate station [dai trung gian] (VTG in Quang Ngai), and the matter of ensuring the transmission of the content of command guidance from Central to the sectors in Nam Bo was more accurate and timely.

Also in 1949 the cryptographic organization of the Saigon-Cholon Special Sector came into being to ensure the command requirements of units active within Saigon.

****

Along with the move to develop and improve cryptographic systems, research into building a theory of cryptographic technique and distilling the essence of experience in the use of technique received highest attention. The Cryptographic Bureau began to compile documents to teach theory concerning the science of cryptography and documents to guide in practice. Experience in raising the level of technique use, in searching out erroneously encrypted values [ky hieu] and groups, was compiled to produce widely publicized documents for people directly involved in doing cryptography--compiled by the cryptographic organizations of the Intersectors, the divisions [dai doan], and the General Staff.

At the end of 1948, a set of cryptographic theory documents with the basic essence, a first system, was compiled. The book *Fundamentals of Cryptography* [Mat ma dai cuong],[11] with Cde Hoang Thanh (Cryptographic Bureau cadre) the chief author, was published by the Ministry of National Defense. It comprised five parts and thirteen chapters.

Cde Ta Quang Buu, Minister of National Defense, wrote the foreword of the book. It included this passage: "Although the laws of cryptography are universal laws, each and every nation must comply with the rules of cryptography of that country, which will differ from another country, because language structure differs from one language to another, or, to say it more scientifically, the frequency of letters and 1. Letter groups are not the same in one country as another. Because of this, the science of cryptography in our nation is still in the research stage.""

In the introduction, the author wrote some unvarnished, sincere thoughts: "When you look at our nation, cryptography is a new branch--it was born with the army and it grows up with the army."

The contents of this book initially dealt with fundamental problems of cryptographic theor--brought out principles of encryption, and basic methods of cryptography.

Cde Brigadier General Chief of the General Staff Hoang Van Thai assessed [this book]: "Mr. Hoang Thanh's book, *Fundamentals of Cryptography*, published in late 1948, was most timely, and it greatly aided the military cadre as well as people specializing in cryptography. Although that book was fundamental, basic, inadequate, compared with today's level and requirements, it helped in no small way those people who needed to use cryptography every day, to have general knowledge of military communications, and it helped people specializing in cryptography to understand methods of applying flexibility and coming up with additional creations."[12]

The book was circulated in command organizations and cryptographic organizations army-wide. The result was to greatly increase the efforts of the Ministry of National Defense and General Staff cryptographic organization.

Also in 1948, the Cryptographic Bureau compiled and promulgated the "Ten Commandments" for the specialty, the first step in laying the foundation for building a system and principles for professional technique for our branch.

At the end of March 1949, on the occasion of summarizing the second "Build the infrastructure, Break the record" emulation, the book *The Use of Cryptosystems* [Cach dung luat mat ma], also authored by Cde Hoang Thanh and published by the Ministry of National Defense, was printed.

The content of the book pointed up the role and importance of cryptography, methods of ensuring secrecy through cryptography, and some essential principles in the use of cryptographic systems.[13]

This was a seminal document, one that fairly well summarized the structure [he thong] and noted the administrative principles in the use of cryptographic systems, because, as the author wrote, "using [cryptographic systems] improperly is like, as it is said, 'moving sand like a sandcrab': You knock yourself out and lose time, but you don't get anywhere."

Cde Hoang Van Thai wrote in the foreword, and for dissemination in the entire branch:

"With a style that is humorous, simple, and colloquial, the author of this volume honestly hopes that, in day-to-day combat, people using cryptographic systems with new experiences and new creations will add on, so that our national cryptography may be enriched."[14]

The book *Fundamentals of Cryptanalysis* [Ma tham dai cuong], compiled by the Cryptographic Bureau in 1949, fell within the plan to compile books on cryptographic deliberations, but not published, however, for it was also a document "to generate sensitivity toward cryptanalysis and help people making cryptographic systems to find ways of avoiding mistakes normally encountered and to progress toward the work of searching out enemy systems."

Relying on the books and study materials in HQ's training classes, in 1948 the Intersector 4 Cryptographic Section compiled and produced a manual used for regimental and battalion cryptographic organizations. It was called *The A.B.C.'s of Cryptography* [Mat ma thuong thuc].

In 1949, until the beginning of 1950, we launched a series of campaigns in the North: Song Lo (April 1949), Song Thao (May 1949), Le Loi (Hoa Binh, November 1949), Le Hong Phong I (Northwest, February 1950). The Cryptographic Bureau designated cadre and personnel to participate in serving the campaigns and organized liaison nets to ensure campaign command with the left bank and right bank fronts of the Song Da. Cryptographic cadre and personnel of the 174th and 209th regiments, etc., for the first time served command in combined operations, incessantly pursuing and striking the enemy, sometimes on the march, sometimes at work, lacking experience, thus occasionally thrown into a passive mode and at a loss as to what to do. Following the line clearly enunciated at the sixth Central Conference of Cadre, ". . . concentrate cadre, concentrate weapons and means of communication-liaison for units with the mobile strike mission," "tables of organization, training, equipment--all must aim at the objective of carrying out the realization of mobile warfare," the Cryptographic Bureau drew experience promptly and instructed the units in arranging the various types of systems and preparing means of responsiveness to command requirements. The bureau also concentrated research on the improvement of cryptographic systems to respond to the requirements of mobile operations and increased its directives on organizing the cryptography of the main force units in order to adequately perform the mission they had received.

The requirements of the new mission placed upon the army cryptographic branch with respect to such aspects as organization, cadre and personnel, and technique, were large and difficult from the outset. The General Staff plan of assignment for 1949 also laid out concentration of weaponry, cadre, and personnel in the technical specialities for the main force; building of the specialty services from HQ down to Intersectors and main force regiments.

In September 1949, the Cryptographic Bureau opened a training class with the class name, "Dong Thap Muoi," meeting in the village of Dong Dau, Dinh Hoa, Thai Nguyen. Nearly forty students from units in the North went to study. The content of the curriculum was augmented in the areas of politics and the situation involving the new mission. This class had a rather methodical routine for training and close direction, with places to study, live, and work for the students and elements, report cards noting such aspects as morals, capacity for study, and professional technique, and record of merits and demerits of each student. Upon completion of the class, the students received graduation certificates from the Ministry of National Defense (MND).

In 1949, because of the expansion of the resistance, the volume of secret messages of command and direction increased rapidly. On an average, each month the Cryptographic Bureau of the MND/High Command encrypted and decrypted 1,200 official messages. Liaison nets also expanded greatly. System usage from Central to sector was examined. A few systems were a bit difficult to use, with a high degree of complexity, thus seldom used. Besides the stations in contact before, the Cryptographic bureau also developed and kept contact with MT3; MT2; the Song Thao Front, the Hanoi Resistance Committee; the Nam Bo Resistance Committee; HQ of Sectors 7, 8, and 9; and HQ, Nam Bo.

Station MT3 was a mobile station with responsibility for liaison with Intersectors 1 and 10, the High Command, and the Resistance Committees. The Cryptographic Bureau selected two representatives to go take charge and six unit personnel to help out with this station. Although their principal mission was encrypting and decrypting, the fellows at MT3 still participated in the building of cryptographic organization for a number of regiments and battalions within their sector of assignment.

A cryptographic team regularly provided direct service to General Vo Nguyen Giap in liaison with the sectors in circumstances involving Flash [hoa toc] cryptograms, liaison with the Secretariat of the President and government, the Viet Minh Central Executive Committee [Tong bo Viet Minh], and the Intelligence Directorate [Cuc tinh bao]. The team had two comrades, with encrypting and decrypting their essential mission.

At the beginning of 1949, per directive from the MND-High Command, [we were to] "correct troop organization, aim at making branches [nganh] and army branches [binh chung] lean, light, but of adequate numbers, consistent with the requirements of mobile operations. . . correct command mechanisms [bo may chi huy] and the specialty branches [nganh], train cadre and personnel from the standpoint of service, fostering professional specialization, and a number of other matters." The Sixth Army-wide cryptographic conference convened from 20-27 June 1949 in Viet Bac comprised as delegates the cadre in charge of cryptography in units in the North and Intersector 5. (Nam Bo and units in distant theaters were unable to come.)

The Conference reviewed the situation involving the task of the army cryptographic branch from 19 December 1946 to May 1949, and went deeply into review of each aspect of the task: building organization, developing cadre and personnel, research into expanding cryptographic technique, and organizing to accomplish the specialty task. Cryptographic organizations from each Intersector reported on their own reviews prior to the conference, in order to exchange training and extract collective experience.

The conference settled on the appropriate duration of training and resolved the direction of the new mission with respect to building an army cryptographic organizational system from top to bottom, unified in thought, organization, and professional technique, in order to have sufficient good outcomes in each mission in the new stage. The conference produced emulation criteria for units in the two years, 1949-50, with respect to development of cadre and personnel, research in the production of cryptographic materials, and emulation to guarantee secrecy, speed, and accuracy in the task of encrypting and decrypting.

In this conference, cryptographic systems researched and produced by the Intersectors and central were likewise reported and presented, aiming at exchange of training and drawing of experience.

The conference was honored to greet the comrade Chief of the General Staff, arriving to visit and speak. Carrying out instructions from the Chief of the General Staff, the conference settled a number of problems:

- Correcting organization

- Improving technique

- Developing cadre, etc.

As to organization:

Made progress in realizing a new, set system of organization from above; speeded up the task of expanding organization down to companies, because, as the demands of combat were expanding in strength and complexity, liaison needed to become quicker and more secret, especially in operational orders, for which the expansion of cryptography to the companies had become an essential ingredient.

As to the specialty:

Had to strengthen and consolidate the task of encrypting and decrypting in order to realize the slogan,"swift, secret, and careful." To carry out the summarization of encryption-decryption experience in order to publicize it for the units. To unify the methods of the task, the way of working; to promote specialty rules of conduct.

As to production and research:

To continue to research and produce per the resolution of the fifth conference... Both at Central cryptographic and cryptographic at the Intersectors there had to be produced a new type of system in order to break away from the systems in use. The Central Cryptographic Bureau type of system had to be in place by the end of 1950.

As to training:

Start refresher classes and develop new people. Build a base for a cryptanalysis branch.

As to the cadre task:

Pay attention to speeding up the task of developing cryptographic cadre, especially cadre with capability, education, and a specialty technical aptitude in order to serve in the stage of preparing for the general counteroffensive. Getting cryptographic personnel specialized: this point meant that, in order to serve the branch, cryptographers would have to shun unsettled thinking, become more innovative, with the capability of understanding. Thus it would be officially recognized that the cryptographic branch is a specialty branch, just as are other branches.

The conference sent a message to President Ho Chi Minh and the General Commander in Chief, and, at the same time to the Intersector headquarters, pledging single-minded nonstop emulation to build the army cryptographic branch big and strong, so that every mission in the new stage would be well accomplished.

This was a significant conference in the process of building the branch. The conference unified a correct outlook toward the building of organization and the expanding of professional technique; it resolved in a relatively and adequately concrete way those tasks which were immediate as opposed to those previously encountered, created favorable circumstances for strengthening organization and realizing directives concerning professional specialization, unified technique branch-wide, changed quickly to meet the army's requirements for building and fighting in the new expansion.

In August 1949, the 308th Division [dai doan] – the first main force division of our army, with the epithet "Vanguard Division," was formed. The division Cryptographic Section and the cryptographic teams of the 36th, 88th and 102nd regiments appeared, with very opportune reinforcements from the Cryptographic Bureau of the MND/High Command as far as organization, cadre, setting up liaison nets, etc. Cde Hoang Hong Hy, cadre from the General Staff Cryptographic Bureau, was assigned as chief of the 308th Division's Cryptographic Section.

In March 1950 the Cryptographic Section of the 304th Division was formed, with cryptographic teams in the 9th, 66th, and 57th regiments. Cde Ngo Duc Tri, cadre of the Intersector 4 Cryptographic Section was decided upon as chief of the division cryptographic section.

In December 1950 the Cryptographic Section of the 312th Division was formed, with cryptographic teams in the 209th, 141st, and 165th regiments. Cde Nguyen Thanh Mai, cadre of the Intersector 1 Cryptographic Section, was assigned as section chief.

Based upon the previous regimental cryptographic organizations, when consolidated to establish division cryptographic there were reinforcing conditions that took us another step in organization and the professional task.

****

In order to guide the Resistance in its powerful spread, in June 1950 the Standing Committee of Central Party issued a decree to straighten out the organization of MND/High Command and to create the mechanisms for Party direction. Per this decree: The High Command [Bo Tong tu lenh] would comprise the General Staff, the General Political Directorate, and the General Directorate of Supply. High Command organizations were established. "These changes were quite significant in making military mechanisms more concentrated and responsive."

In June 1950, the MND Cryptographic Bureau and the Cryptographic Bureau of the General Staff were merged to form the General Staff Cryptographic Bureau under the direct guidance of the Chief of the General Staff. Together with the specialist organizations in the General Staff, such as Military Intelligence, Ordnance, Militia, Communication-Liaison, Engineers, Quartermaster, etc. the cryptographic organization of HQ "was increased in organization and readjusted in responsibilities in accordance with the requirements and guidance of the liberation struggle expanding into a new stage."[15]

The Cryptographic Bureau of the General Staff was organized in four sections:

- The Research, Production, and Cryptanalysis Section

- The Training and Control Section

- The Encrypting and Decrypting Section

- The Clerical [van thu] Section

In the Encrypting and Decrypting Section, there was an element to watch closely the mission of encrypting and decrypting with the nets and places in the South, one for encrypting and decrypting with sectors in the North, and one for encrypting and decrypting in the service of the campaigns.

The chief of the General Staff Cryptographic Bureau was Cde Hoang Van Dong. The Bureau "continued to correct its organization, researched improvements in the army system of cryptography consistent with the operational mission; unified the organizational system from Central down to basic elements; developed cryptographic cadre and personnel to supply to the units and theaters of war and supplied organizations outside of the army, such as elements abroad; continued to mobilize the cryptographic system production emulation campaign; strengthened the cryptographic base in Nam Bo; and expanded the cryptographic base in Laos-Cambodia."

The bureau had a [Party] cell [chi bo], each section had a Party team [to Dang], and the number of Party members was much larger than previously. During this period the Cryptographic Bureau was augmented by Cde Le Thanh Hai as bureau political commissar, concerned both with Party tasks and political tasks, the bureau having many worker activities-- growth, education and sports – and was a participating unit in the vigorous emulation campaign among General Staff organizations.

There were separate cryptographic elements in the organizations of the General Political Directorate,General Directorate of Supply, and Directorate of Military Intelligence,[16] but they were under the technical and professional direction of the General Staff Cryptographic Bureau.

In 1950 the Main Military Committee appointed Cde Nguyen Chi Thanh, member of the Central Party and concurrently head of the General Political Directorate, to personally lead the army cryptographic branch,while at the same time a number of Party Cadre were posted to the army cryptographic branch. Among these cadre were those of regimental level, some of district committee member level or of party cell committee level.[17]

The work of increasing the number of cadre and merging the MND and General Staff cryptographic organizations, creating favorable conditions for building a strong and stable cryptographic bureau, advanced us well in performing mission responsibilities.

In July 1950, the Seventh Army-wide Cryptographic Conference reviewed and drew experience from the realization of the decisions of the sixth branch conference, continuing to issue direction for the army cryptographic task in the approaching stage. This conference pulled together a number of large problems, such as emulation in cryptographic system development, raising the level of cryptographic technique; speeding up training in raising the level of combat mindedness and the level of political thought on the part of soldier-cryppies; unifying guidance along the vertical axis of the cryptographic task in the army.

It can be said that, by the Seventh Army-wide Cryptographic Conference, the cryptographic branch had built a system from Central down to basic elements. As for responsibilities and the mission of the cryptographic organizations at the various levels, those had also been clearly settled: The General Staff Cryptographic Bureau had the mission of encrypting and decrypting, researching cryptographic techniques (codes and ciphers) and supplying cryptographic materials to the fifty-eight units, training new personnel, cryptanalysis, and guidance and control over the specialty task of the branch. The cryptographic sections of the Intersectors and divisions had essentially two missions, encrypting and decrypting telegrams; researching and producing cryptographic systems [luat] to supply to directly subordinate units; and developing new personnel when delegated from higher echelon.

In order to overcome the obstacle posed by lack of cryptographic cadre and personnel and speedily raise the level of cryptographic cadre and organizations, and with General Staff concurrence, in 1950 the General Staff Cryptographic Bureau organized four consolidated development classes. Two of these, "Le Hong Phong" and "Le Lai," to develop cryptographic cadre, opened in the Ban Co jungle, Yen Thong village, Dinh Hoa district of Thai Nguyen. Cde Bureau Chief Hoang Van Dong personally took charge and taught, together with comrade instructors Ho Quang Chinh and Le Van Bang.

The [other] two classes, "Chi Lang" and "Lam Son," to develop personnel, were under, and taught by, Cdes Tran Dzuy Trong and Le Dinh Y.

The total number of students in these classes was 200 comrades, some being old soldiers, some recruits, some transferred from the Ground Forces School, or party cadre from the Party and Government organizations assigned to go study.

The curriculum content was fairly comprehensive in layout, with respect to political matters, military matters, technique, and professionalism, comprising both theory and practice. Before studying the cryptographic specialty, students studied political matters, to thoroughly grasp the mission situation and establish the thinking of volunteerism in long-term service in the cryptographic branch, after which an appreciation of the value of the organization would result in a formal determination to study. Each of the four classes established a party cell; each comprised party members as instructors and party members as students.

One day early in May 1950 there occurred a great and unexpected honor for the cadre-students in the Le Lai class: Uncle Ho arrived for a visit en route to his task. From the night before the day Uncle promised to drop by and visit, until 0500 the day after, absolute secrecy had to be maintained. From early morning, all of the class assembled, waiting to meet Uncle. The school was in a jungle area at the village of Yen Thong, Dinh Hoa district, Thai Nguyen. Upon seeing Uncle coming by horseback from a distance, everybody, young men and women, stood up and raised the shout, "Long live President Ho!

Long live President Ho!" The shouting rang throughout the jungle. Dismounting, Uncle turned and said to Cde Ho Quang Chinh, who was in charge of the class and had come out to meet Uncle, "I cautioned you, young fellow, to keep this a secret! What in the world is this?" Uncle then went to the place where the young people were and shook hands, saying "No more shouting." Seventy young people surrounded Uncle, like his little nieces and nephews, to meet him. Cheerfully waving his hand for each person to be seated, Uncle's bright eyes and gentle disposition took in the entire class at one time, Uncle asking, "Where are the girls and boys of Sector 3? Sector 4? Viet Bac? Sector 10? They must be around here." Hands were raised in turn. Uncle patted the head of one comrade with malaria, who had lost almost all of his hair, and asked, "Are you still feverish, son?" The comrade student was deeply touched: "Dear Uncle, we got to meet you: now the malaria is gone, sir." Speaking to the class, Uncle recommended along these lines: Our resistance against France grows more victorious, our army grows larger, but still there are many difficulties and hardships to be met. Uncle and the Central comrades work overtime, but we must also go into the jungle to gather bamboo shoots in order to improve. It is good that you young people grow like that. If we lack vegetables we can gather banana buds and banana flowers in order to improve, and make salad, but not have much to eat. As to the mission of studying and the task, Uncle instructed, "Cryptography is a secret task, tremendously significant. It was essential that the General Staff open classes like this. You young women and men have the trust of the Party and the High Command--you must study well and work well. Cryptography must be secret, swift, and accurate. Cryptographers must be security conscious and of one mind. I greet you all and wish you well."

Because he was busy with much work, Uncle only visited the class for a few minutes, but his words and his sentiments of concern for the cryptographic branch were truly unlimited and profound. From that time, three of his words were inscribed--"secret, swift, accurate"-- to become the guidelines for the task of the cryptographic branch, and the word "single-minded" became the sentiment and way of life for the cadre and personnel in the entire branch guiding and assisting each other in accomplishing the mission and building a tradition for the army cryptographic branch that was fully satisfying in every respect. Each person increased his efforts to study, strove for improvement in training, and to raise the level of professional capability, appropriate to the solicitude and trust of the Party and Uncle.

Also from 1950, the task of developing and building the cadre ranks and regulating the selection of people to enter the branch become tighter each day, with special attention to choosing people from a working class background: the selection process for persons entering the branch was carried out with caution. Cde Nguyen Chi Thanh had important instructions concerning the political task, ideology, organization, technique and conduct of work vis-a-vis the army cryptographic branch. The General Political Directorate instructed the [Party] executive committees to select their very best for cryptographic work and diligently organize to cultivate cryptographic cadre and personnel who had not yet entered the Party to become party members. By the end of 1950, almost all cryptographic cadre and personnel in the army were party members.

Besides the cadre development classes at HQ and the Intersectors, on-the-job organization and development continued to be carried out, according to the requirements of cryptographic communication nets [mang thong tin bang mat ma], such as in the regions temporarily in enemy hands, and the distant theaters, such as Trung Bo, Nam Bo, etc.

In league with the above activities, the branch also commenced the implementation of a tighter system of administration, promotion, and employment of cadre and personnel in the branch. A number of systems of commendation and reward and fostering came to the attention of the upper echelon.

The task of developing army cryptographic cadre and personnel in the years at the beginning of the resistance involved many efforts to respond to mission requirements.

EMULATION TO BRING INTO PLAY AND IMPROVE THE LEVEL OF CRYPTOGRAPHIC TECHNIQUE

From 1950, the resistance was stronger, but larger-scale campaigns were opening, one after another. The cryptographic organizations and cryptographic warriors on assignment all over, in the provinces and theaters, had overcome many tough situations in order to meet the wide-ranging requirement to ensure command by cryptography in battle. And also, through the practice of serving in battle, many weaknesses in the type of technique being employed were exposed and needs had to be instantly taken care of.

Vis-a-vis our cryptographic branch, the French colonialists had organized many cryptanalytic organizations comprising many experienced people, while at the same time expanding their espionage agents to plot to exploit our secrets, from central organizations down to the sectors. In one dash against lines of communication in Nam Bo, we collected a quantity of enemy cryptanalytic documents, among which were cases in which our messages had been broken out, etc.

Marking well the words of Uncle: "If we know the enemy clearly, then we win. If the enemy knows us clearly, then we lose, so we must ensure secrecy. . .," the army cryptographic branch constantly set a high level of vigilance, never ceased to improve and raise the level of technical secrecy, while the communications troops "regularly change cryptography and station schedules, we militarize the radio station and cryptographic organizations" per the instructions of the Central Party Standing Committee.

At the Seventh Army-Wide Cryptographic Conference (1950), the role of the technique research task was defined more concretely in the overall task of cryptography, generally speaking. The responsibility and authority of the technique research organizations were decided upon as follows:

"The Research and Development Team in the Research and Training Section has the responsibility and authority for research and development of cryptographic systems, and for handing long-life systems to the Intersectors, divisions, and directly subordinate units, the organizations in the High Command, and delegations, and for monitoring and remedying the cryptographic situation nationwide.

"The R&D Team in the Campaign Cryptographic Section (subordinate to the Cryptographic Bureau) has the responsibility of R&D on systems to serve campaigns, essentially for the main force divisions and regiments, and for monitoring and remedying the cryptographic situation in the sphere of main force units engaged inaction."

In the Intersectors, "The Cryptographic Section has one R&D cadre, with the responsibility for R&D, distributing systems to the regiments and provincial units, battalions, and district units, and monitoring and remedying the cryptographic system situation in units served.

"Creations of cryptographic material must be sent up to the Central R&D organizations for examination, remedy, and drawing experience. Whenever the R&D task is adequately taken care of with cadre and personnel and means of making cryptographic systems for the regiments and provincial units, the Intersector will not have the above responsibility and authority.

". . . In essential cases, the division cryptographic section may make and give out codes and secret signals [am hieu] for essential units and report to the Bureau. . . ."

Warmly responding in years past to Uncle Ho's call for patriotic emulation and implementing the order "Train the Army, Achieve a Feat," from the High Command, the cryptographic branch officially mobilized a "high" in the emulation movement, "Bring into Play Innovation and Improvement of Technique," by practical slogans, specifically, "Each Cryppie a Compiler [of cryptographic systems]." In the competition to develop cryptographic systems, there were scores of the best selected by the Intersector Cryptographic Sections and sent to the Cryptographic Bureau for examination. These systems comprised many different forms [the]: three-element [the], four-, five-, and six-element--mostly four-element and three-element.

Four-element systems sent to the Bureau from the units for reporting and examination were used in the units in all shapes and forms. Four-element chart systems were used in nearly all conceivable forms. Depending on method, the four-element could be used for encipherment perpendicular combined with right angle or encipherment entirely by right angle.

At the end of 1950, the General Staff Cryptographic Bureau, i.e., two technique research cadre (comrades Dzung and Con), using as the model a chart system with a method of breaking down and combining [words in] two parts, invented in Intersector 4 in 1949, applied it in a four-element form, and produced the type of system of preeminence, known as ZC-4 [Vietnamese DzC-4: Dzung and Con's 4-element system?].

Code chart DzC-4 used a complete square for encryption, using substitution by 4 encrypted values [ky hieu ma]. If the number of plain units was not large, one could use chart sizes 13x26 up to 26x26, according to circumstances and conditions of use.

System DzC-4 brought into play many strengths and showed creativity, as a type of system with high value from the standpoint of protecting secrecy and also of value in practical use, for it broke up and combined syllables simply, logically, neatly, and purely, with a basis that was both scientific and fresh. Encrypting and decrypting were quick and simple, and the capacity for accuracy was high. Messages were shorter, compared with encryption by other types of system; speed increased in converting the contents of secret messages. The system could be used with many types of key strips, a favorable consideration in production and use. The form of substitution was uniform for all plain units in the chart, the ratio between the two parts of a syllable in composition being 115/145. This satisfactorily resolved the balance in frequency between the two components.

The DzC-4 code chart clearly took us a step forward, marking the growth of the code chart research task of the Vietnamese cryptographic branch.

The DzC-4 form of system afterward spread rapidly in application among the units. Types of systems similar to DzC-4 were Intersector 5's 1H1-54 system, Nam Bo's HCM [Ho Chi Minh?]-3-53 (used in 1953), a system of the 325th Division (in the years 1952-1954), Intersector 4's Hoa Binh [PEACE] system (used in 1954),etc.

The three-element system was also produced in many types: by 1950 it was being used widely in Nam Bo.

Task standing operating procedures, principles, and tables of organization with respect to professional decisions of the branch received close and thorough attention from the responsible organizations at the various levels from day one. From the "Ten Commandments of the Specialist Task" through the realities of guidance and use, the Cryptographic Bureau came up with the "Thirty-four Commandments of the Specialist Task," dealing concretely with the problems of protecting cryptographic secrecy and the internal aspects of the task, with respect to the work practices and relationships of cryptographic cadre and personnel. The Cryptographic Sections got element and individual responsibilities nailed down. In the Bureau and a number of Intersectors and divisions, specialization was realized net by net, each person receiving a fixed, set number of systems, the person using them to be responsible for following the observations of strong and weak points noted in use of the cryptographic materials, both to complement the construction of systems that became richer with each passing day, and to avoid technical errors.

The task of technical direction and use had positive measures, mainly in not letting the enemy have the time and circumstances to implement trickery in order to penetrate our cryptographic systems.

The emulation movement made systems, researched, and immediately produced systems on the spot as a result, but there were systems constantly guaranteed with strict regulation with respect to period of use, replacement of key strips [bang khoa], increasing the types of reserve systems to prepare for prompt replacement in a time of necessity or a sudden task, etc.

From 1950 on, systematic review and assessment of procedures and technique was constantly maintained in every unit. The rules and processes of encryption and decryption never ceased to be perfected and made uniform in the branch. The chain-link method of operation was applied, involving mind-set, hands, eyes, and nervous reflexes on the part of each person encrypting or decrypting. The method of reading code values twice-over was improved by a single reading, the effect being to cut the time and eliminate mistakes in hearing and reading letters of the alphabet [chu cai]. Many comrade cryptographers were highly polished, becoming skilled cryptographers with a proficient level of professional technique and enhanced productivity, capable of meeting command and control requirements.

THE CRYPTOGRAPHIC TASK IN THE AUTUMN-WINTER BORDER CAMPAIGN OF 1950

By 1950, the international and domestic situations had evolved favorably for us.

In August 1950 the Central Party Standing Committee decided to open the Border Campaign (still known as the Cao [Bang] - Bac [Kan] - Lang [Son] Campaign): "The campaign requirements are to annihilate a vital, important enemy element; to liberate the northern border region of our nation; to restrict the scope of the enemy's occupancy; to enlarge and consolidate the Viet Bac base area; to progress toward seizing the strategic initiative in the main theater."[18]

After the politico-military conference of 24-25 August 1950, permeated with the determination of the campaign Party committee, the comrade Chief of the General Staff decided to set up a forward staff organization of the High Command to perform the functions of a campaign staff [Bo tham muu chien dich], comprising operations, military intelligence, cryptographic, communications, and administrative management sections.

The campaign took place in an extensive area of mountainous jungle, thus ensuring that communications-liaison and cryptography would be a serious problem. In accordance with a directive from the Chief of the General Staff, the Cryptographic Bureau urgently prepared the organization to ensure a cryptonet. The bureau convened a conference of cadre in charge of cryptographic organizations from the units participating in the campaign, thoroughly grasping mission requirements, unifying a plan to align cadre and personnel forces and setup a cryptonet, and giving direction in the use of technique and methods of organizing for work in the campaign.

The Campaign Forward Cryptographic Section and the mobile cryptographic teams serving reconnaissance and command of the General Staff organs were organized so as to align forces for the expanded mission. The cryptographic forces of the participating units also were mobilized to a high level, comprising

The 308th Division Cryptographic Section and the cryptographic teams of the 36th, 102nd, and 88th regiments.

The Intersector Viet Bac Cryptographic Section and the cryptographic teams of the Cao Bang and Lang Son Provincial Units.

The General Supply Directorate Forward Cryptographic Section.

The cryptographic teams of the 174th and 209th Independent Regiments.

Mobile cryptographic teams.

On 1 September 1950, the Campaign Forward Cryptographic Section, under Cde Nguyen Chanh Can, together with the campaign staff organizations from HQ, set out for the front. On 4 September in Na Lan (Cao Bang), cryptographic cadre and personnel constructed places to mess, quarter, and work, as a matter of urgency, while serving command of the troops maneuvering to consolidate and transporting rear services materiel for the campaign.

This was a campaign in which main force troops were first concentrated for large operations: from the beginning to the end of 1950 we had the artillery and engineer branches combined in combat with the infantry--a lengthy campaign, with many large engagements following large-scale forms of operations, compared to previous campaigns. The task of ensuring cryptographic technique in the service of command produced new requirements, complex and more urgent. The campaign cryptographic organizations and the units alike lacked cadre and personnel, professional knowledge and means, and experience in organizing to implement the tasks of a large-scale campaign. However, throughout the campaign, cryptographic cadre and personnel overcame each obstacle to meet the requirements of command and control. Many operational orders were encrypted and decrypted quickly, accurately, and in a timely manner. Cryptographic cadre and soldiers had the honor of handling messages containing Uncle's recommendations to the troop units concerning the Cao-Bac-Lang campaign: ". . . The Cao-Bac-Lang campaign is very important. We must resolve to win the battle; soldiers on this front must be determined, 100 percent valiant; soldiers in the sectors and other fronts must strive to emulate in killing rebels in record numbers, so we can wipe out the enemy, pin down the enemy, not permit him to reinforce the Cao-Bac-Lang front."

On 16 September the battle of Dong Khe raised the curtain for the campaign.

In the process of carrying out the campaign, messages of operational direction from the comrade Commander-in-Chief and the comrade Chief of the General Staff, along with many messages from the command comrades at other high echelons, were quickly encrypted and sent along to the divisions and regiments, the units participating directly in the campaign and the cooperating theaters.

As a special note, on 30 September and 1 October, when we knew, thanks to enemy messages we intercepted, that the enemy was carrying out a plan to withdraw from Cao Bang and reoccupy Dong Khe, the activity of the cryptographic organizations became more urgent in serving the campaign command's designation of forces to strike the enemy.

Each day, on the average, the Forward Area Cryptographic Section encrypted and decrypted over 100 official messages. Thirty-seven continuous days and nights of decisive combat, and the operational situation between ourselves and the enemy was tense, daily message volume was therefore rather large--the task of encrypting and decrypting messages and handling requirements messages had, in truth, to be quick and exact, especially in the matter of ensuring thorough grasp and accuracy of the campaign CP decision to wipe out two groups, those of [Colonel] LePage and [Colonel] Charton. At 2330 hrs on 5 October 1950, the Cryptographic Section received Order No. 8 signed by the commander and concurrent political commissar of the campaign, Gen. Vo Nguyen Giap, going to the 308th Division, the 174th and 209th Regiments, and the provincial units of Lang Son and Cao Bang, with the precedence indicated on the order to be the highest: "Flash," executing the decision to wipe out the enemy's LePage and Charton groups. The parts concerning the situation and decision were quite clear.

I. Situation and Estimation:

1. The LePage group is currently in Coc Xa and along the Quang Liet mountain range. They are set to seize hill 477 in order to make contact with the Charton group and simultaneously have an element of a Legion parachute battalion go over and occupy Quy Chau hill.

2. The Charton group may follow the Mong Xa road to the Ban Cao-Lan Hai section, then have an element go up and occupy hill 477 to link up with LePage's forces to fight us and save the situation.

3. Tomorrow enemy aircraft will heavily bomb from Coc Xa to Pac Bo, especially hill 533, to provide cover and support and to open the road for the two above groups to pull back to That Khe.

4. Tonight we attack and strive to wipe out the LePage bunch in the Quang Liet region. The combat may extend into tomorrow.

II. Decision of the Campaign CP:

1. Take advantage of the time to wipe out the LePage group before the Charton group closes.

2. Harass, wear down and wipe out in detail the Charton group to create conditions so that after the elimination of the LePage bunch we can consolidate forces to mobilize for a total wipe-out. . . .

46

The order went on to bring out the specific mission of units in the Quang Liet region, reserve units controlling and threatening Khau Luong and Keo Ai, and the 209th and 174th regiments, diligently controlling and threatening, actively surrounding, pursuing, and wiping out the enemy.

Encrypted, the order resulted in a twenty-part message. The cryptographic organizations and the communication organizations in the campaign meshed together to ensure transmission of the order to the units with timeliness, accuracy, and secrecy.

One illustration of the agile spirit displayed in ensuring command by cryptographic means in the campaign is the initial case in which the cryptographic organization and the communications organization set up the "station flap" [op dai] method, a method of direct exchange between the campaign command and the command of the 308th Division via cryptographic system and radio. This was during the opening of the Border Campaign-- our forces had assaulted Dong Khe and the combat situation was becoming critical. Our forces and those of the enemy had become exhausted by mid-day. The troops alternated between movement and attack of the enemy. Combat orders from the division commander at that time had to go for the most part by cryptographic system and radio. Division Commander Vuong Thua Vu called cryptographic and radio up to the CP so that he could personally command the regiments via cryptographic and radio. In accordance with orders from the comrade chief of the division cryptographic section, Vu Van Can, Cdes Phan Tien and Linh Son were selected to go. Under "station flap," if cryptographic had two people, then one would encipher or decipher and one write, or, [in the case of] radio, there would be one operator and two assistants to crank the generator (Ragonout). A command person would sit beside the cryppie and personally write or read the message for enciphering. Once the message was encrypted, cryptographic would hand it to radio to send on. If it were an incoming message, the cryppie would receive the cipher message from the radio receiver and, once it was decrypted, pass it on to the command person to examine. Cipher messages under these circumstances were usually under ten groups and at urgent precedence, but usually "station flap" meant intense labor by the cryptographers and radio station personnel, having to mobilize to a peak level both for productivity and quality of encrypting and decrypting, when called on – it was not only a matter of skill in one's specialty, but also a matter of having to settle matters fast.

Many times the comrade commander of the 308th Division, Vuong Thua Vu, and the commander of the 209th Independent Regiment, Le Trong Tan, called cryptographic and radio up to the CP to serve the command function directly. There were also times when comrade General Vo Nguyen Giap contacted Comrade Vuong Thua Vu and Cao Van Khanh via "station flap." Initially, because of a lack of experience, the procedure for encrypting a message called for complete compliance with the rule "indicate the strip, indicate the key," etc. But through the experience of timeliness requirements and the length of cryptograms, combined with the characteristics of the type of technique of the KTA chart-code [luat bang KTA], the comrade cryppies developed procedures so that when in "station flap," then they would settle on one type of key so as not to have to indicate the

47

key, [would] shorten the cryptogram, reduce the formalities, but still maintain secrecy and accuracy, thus speeding up the time for enciphering and deciphering.

The majority of "station flap" messages attained high results--information passed via cryptographic and radio took the shortest route. Command comrades Vuong Thua Vu and Cao Van Khanh saw the cryppies working under extreme pressure and not only encouraged their cryptographic spirit but were concerned to support them in all matters.

Generally speaking, "station flap" liaison brought into play a high level of action, but there were also instances in which requirements were not met because the cryptographic system had not been set up scientifically – sensibly – the content of the system was still deficient, and the plain/encrypted ratio not high, making encryption and decryption untimely.

Along with the cryptographic organizations in the Cao-Bac-Lang campaign, the cryptographic organs in the cooperating theaters and the fronts behind the enemy served the unit command task well. In Northwest, the cryptographic team of the 148th and 165th regiments served in the plan to distract and strike the enemy, liberating the entire left bank of the Red River within Lao Cai province and part of the right bank of the Red River up to Sa Pa. The cryptographic team of the 246th regiment (in the Trung Dzu theater) ensured unit command in striking paratroopers in Thai Nguyen. The cryptographic organizations of the 304th Division served in actions behind the backs of the enemy, coordinated with regional troops and guerrilla militia in Intersector 3. The cryptographic organization in Binh-Tri-Thien theater ensured command in combat in the "Phan Dinh Phung" campaign, striking the enemy in Quang Binh and Quang Tri and hitting communications on the Hue-Da Nang road. In the Intersector 5 theater, the cryptographic team of the 108th Regiment participated in the "Hoang Dzieu" campaign in northern Quang Nam and the cryptographic team of the 803rd regiment served command in the Khanh Hoa front. In Nam Bo, cryptographic of the units served command in the campaigns of Tra Vinh (Sector 7), Ben Cat (Sector 8) and Long Chau Hau (Sector 9).

****

48

On 22 October 1950, the Border Campaign was victorious. We killed and captured 8,000 of the enemy and liberated a stretch of the border 750 km long, with 350,000 people. "The Border Campaign was our first large-scale offensive campaign--a mobile striking campaign, hitting and eliminating the enemy in a first-class manner, achieving the highest combat results for our army and our people."*

Exactly as Uncle Ho said, "The victory in Cao-Bac-Lang is a victory shared by the soldiers of the entire nation." The cryptographic cadre and soldiers played a fitting part of their own in that victory.

The campaign cryptographic organizations diligently surmounted obstacles in developing and arranging a cryptographic network that was timely and accurate and that ensured the task of encryption and decryption with a high volume of message traffic, in combat conditions of continuous movement, requiring a high degree of timeliness. Each unit accomplished its mission; there were no major errors to interfere with command guidance. Summarizing the campaign, Uncle Ho recalled some points that needed corrective action, among them the matter of ensuring secrecy. At the campaign recapitulation conference (27 November 1950), Cde Commander-in-Chief Vo Nguyen Giap stated clearly: "People in command are not paying sufficient attention to the organization and use of radio and cryptographic--radio and cryptographic are being placed distant from command personnel (notwithstanding the directives for command levels and Intersectors that radio and cryptographic must be placed near command personnel). But these directives are not being adequately grasped and carried out in the units, so we have orders by radio not reaching command personnel until a day later."

Through service in the campaign, the army cryptographic branch drew valuable experience in organizing to ensure command.

Cryptographic organizations at the various levels had to thoroughly grasp the mission situation, as to campaign and combat intentions; had to prepare the types of cryptographic materials and plan for employment accordingly; had to ensure that the resources for encrypting and decrypting were concentrated on contents that met the high estimates of timeliness; constantly strove for close cooperation with the operations organizations and the communications organizations in the process of campaign preparation and implementation, aiming at ensuring firm, solid grasp continuously for command, especially when the unfolding situation is urgent. The process of preparation must have anticipated developing situations, must have prepared conditions to ensure a positive, proactive method when the requirement is levied.

---

* "When the smoke cleared, the French had suffered their greatest colonial defeat since Montcalm . . . died at Quebec." Bernard B. Fall, *Street Without Joy: Indochina at War, 1946-1954*. Harrisburg, PA: The Stackpole Company, 1961, 28--Tr./Ed.

Also through the Border Campaign we came to see clearly that we had to have thoroughly penetrating creativity in organizational guidance to execute the task of encrypting and decrypting, following the flow of outgoing and incoming messages in an exact manner. When the scale of the campaign and the form of operations changed, then cryptographic organizations had to take the initiative in concert with radio and with upper and lower [echelons] to ensure that command requirements were fully grasped in every situation; had to quickly come up with and put forward ideas for the command people when impediments were encountered in the specialty mission and in task relationships, in order to speedily discern methods of solution and avoid situations of being late and causing adverse influences for combat.

From the standpoint of technique, there had to be types of systems suitable to meet the requirements of mobile operations, compact with respect to form and quantity, adequate with respect to compilation of code content, reliable with respect to degree of security, etc.

In the task of deception and the timely breaking out of the contents of enemy messages, the cryptographic organizations had notable contributions. This was a new task in serving to ensure victory for the campaign. And the enemy also had to recognize their failure in this respect.

Through the Border Campaign, the cryptographic branch affirmed its maturity, meeting the ever more demanding requirements of the combat mission outstandingly, serving the army in increasingly larger and successive campaigns.

The significance of the Border Campaign victory is very great in the history of the resistance of our people against the French colonialists. The gates of the northern border were opened, to provide access to democratic and socialist nations and create favorable conditions to win international assistance from friends everywhere in the five continents. Right after the Border victory, the High Command decided to designate a group of cryptographic cadre to go over to China to study technique and cryptographic professionalism. The group had forty-three comrades with Cde Hoang Van Dong, Chief of the General Staff Cryptographic Bureau, as chief of the group and Cde Nguyen Trieu as political assistant. Cde Nguyen Dzuy Phe was designated to replace Cde Hoang Van Dong in charge of the Cryptographic Bureau. The group studied in China until May 1951, then returned home.

THE CRYPTOGRAPHIC TASK IN THE VOLUNTEER UNITS FIGHTING ON LAOTIAN AND CAMBODIAN SOIL

When the sounds of gunfire erupted nationwide, the Indochinese Communist Party Central directed establishment of the Western Front, consisting of the Cambodian Front and the Laotian Front, to help the Cambodian and Laotian revolutions, together with Viet-Nam's, to implement the resistance against the French colonialists and gain independence. In order to ensure command secrecy for our volunteer forces and the friends, active on the battlefields of Cambodia and Laos, the volunteer units organized a cryptographic system [he thong]. Because communication [giao thong] conditions were difficult and activity was in the heart of the enemy, initially the comrades in the Western Front researched and compiled for themselves cryptographic systems for liaison to ensure command secrecy between the Nam Bo Sector [Xu] Committee, the Overseas Vietnamese Party HQ Special Committee, and the Western Front.

The cryptographic systems at this time were still simple, using one-time substitution: each message was enciphered with its own key, which, once enciphered, was immediately destroyed. The cryptographers used four open-source publications to derive [tao lap] cipher key:

- The 2 August 1945 Declaration of Independence,

- A selection of the literary classic, *Kieu,*

- Dimitrov's 1947 report (in translation),**

- Cde Truong Chinh's "Protracted Struggle But Sure Victory"

In January 1949, the Central Cadre Conference meeting from the 14th to the 18th issued a resolution concerning the military mission, namely, the opening of the Lao-Cambodian Front. Per a directive from the Central Standing Committee, Intersector 4, 5, and Nam Bo were themselves to organize and bring up to strength armed propaganda units to expand activities to establish political bases in the sectors of Laos and Cambodia.

As a result, ensuring command secrecy for the directives on activities to assist the friends in Laos and Cambodia was unified under the cryptographic organizations at Intersectors 4 and 5 and the Nam Bo Command to directly organize and arrange cryptographic nets for the volunteer units going over to be active in aiding the friends.

In Cambodia, the Southwest Liberation Committee had been established in March 1948. By the end of 1949, the base sectors had connected up with one another, and the liberated region of Cambodia had expanded. In April 1950, according to the policy of the Agents' [can su] Committee of the Cambodian Nation-Wide Party, the Khmer Issarak Front was established, creating the Cambodian People's Armed Forces. The Vietnamese Volunteer Army units increased their assistance to the friends in building and in fighting, following instructions from the Central Standing Committee of the Indochinese Communist Party.

** Georgi Dimitrov (1882-1949) – Bulgarian Communist; secretary general of the COMINTERN, 1935-43; premier of Bulgaria, 1946-1949. Reference is presumably to his presentation at the organization meeting of the COMINTERN in September 1947. Tr./ed.

At that time, the cryptographic organizations of the Vietnamese Volunteer Army started to expand in Cambodia. The cryptographic organization in the All-Cambodia Agents' Committee, under the charge of Cde Cuong, concurrently was responsible for cryptography for the Agents' Committee of East Cambodia. Cde Oanh was in charge in the Northwest Agents' Committee, and for the Southwest Agents' Committee, Cde Kha was in charge of cryptography.

In the Volunteer Army battalions, there were one to two cryptographic personnel performing the mission.

The cryptographic organizations of the Volunteer Army served command leadership in executing the missions and tasks of assisting the friends in building bases, building forces, and armed propaganda activities; transporting supplies and weaponry, etc., participating in the development of the friends' resistance forces.

From 1950 on, the cryptographic organization of the Volunteer Army in Cambodia received help from the Nam Bo HQ[19] Cryptographic Section, with respect to people and cryptographic material as well as task experience, and started to receive professional guidance from high echelon cryptographic organizations.

In Laos, in 1948 the task assault units and the Lao-Viet armed propaganda units expanded, active into Sam Neua, Xieng Khoang, and the provinces of Central Laos. In September 1949 the Lower Laos base sector was established. By the beginning of 1950, many resistance base sectors had taken shape, from Upper Laos and Central Laos to Lower Laos. The Lao armed forces were consolidated and expanded further, coordinating combat closely with the Vietnamese armed forces.

The Cryptographic Bureau of the General Staff, cryptographic organizations of Intersectors 10, 4, and 5, and of the Vietnamese volunteer army units served efficiently for command guidance, participating in helping the Laotian revolution expand.

In December 1949, we opened the Song Ma Campaign. The Cryptographic Section of the Northwest Front, cryptographic teams of the 138th and 148th regiments, and other units did a good job of ensuring for our units coordination with the friends' forces wiping out the enemy at Xieng Kho, breaking through the enemy's Song Ma line from Muong Sam to Sop Hao, enlarging the liberated area by more than 2,000 square kilometers, with 10,000 people.

In February 1950, the Lao Patriotic Front (Neo Lao Hak Sat) was established and officially organized the Lao People's Armed Forces in Upper Laos. The Lao Issarak armed propaganda units were zealously active in expanding and consolidating guerrilla bases and liberated areas. At the same time "Westward Ho" [Tay tien] army groups of the People's Army of Viet Nam crossed over to coordinate activity in accordance with an agreement between ourselves and the friends. Based on the mission requirements of each Intersector and the concrete guidance of the General Staff Cryptographic Bureau, the Cryptographic Sections of Intersectors 4, 5, and 10 organized cryptographic elements to go and perform the mission in the Westward Ho units, comprising the cryptographic teams of Group 80 in Sam Neua, Group 81 in Xieng Khoang, Group 83 in Vientiane, Group 101 in Lower Laos, and Group 102 in Central Laos. Also in 1950 we organized a task group to help our Lao friends, with the name Group 100. The Cryptographic Section of Group 100 was established with Comrade Nguyen Ba Zung in charge, and with responsibility for general guidance in the cryptographic task, serving the Vietnamese volunteer units and helping our Lao friends with the cryptographic tasks.

Notes

1. *Military Documents of the Party*. Hanoi: PAVN Publishing House, 1976, vol. 2, 250.

2. Minutes of the Bac Bo [Tonkin] conference of 14 September 1947, quoted in the *History of the General Staff*, 119.

3. Comrade Ta went to Central Party cryptographic; Comrade Nhan went to Ministry of Internal Affairs cryptographic; Comrade Dzuyet went to cryptographic at the Prime Minister's office.

4. Intersector 1 report for 1948.

5. Extract from the report of the Intersector 1 Cryptographic Section at the Sixth Army-wide Cryptographic Conference.

6. Intersector 3 report for 1948.

7. Cde Nguyen Van Dzanh became an augmentee to go down and work as deputy chief of the Intersector Cryptographic Section. Afterward, he was appointed to go down and head the Sector 6 Cryptographic Section.

8. Report of the Cryptographic Bureau at the Sixth Army-Wide Conference.

9. Ibid.

10. *History of the General Staff*.

11. The major parts of "Mat ma dai cuong" were

      1. Fundamentals of cryptography

      2. Substitution

      3. Transposition

      4. Make-up and superencipherment

      5. Conclusion.

The book had an appendix with a list of terms used in the book and their French equivalents. This book of 130 pages was printed in 910 copies (110 copies on good paper).

12. Extract from the introduction to the book *The Use of Cryptosystems*, published by the Ministry of National Defense in 1949, 2-3.

13. Printed 31 March 1949. Contained eight major problems, namely:

      1. Why one must use cryptographic systems

      2. Under what circumstances are cryptographic systems used?

      3. How are cryptographic systems safeguarded?

      4. Safeguarding cryptographic systems with respect to specialist aspects

      5. Requesting repeats when cryptograms cannot be decrypted

6. On the cryptogram log and the duplicate of the secret message

7. Twenty violations of cryptographic rules

8. How are violations to be disciplined?

14. Extract from the introduction to the book *The Use of Cryptosystems*, published by the Ministry of National Defense in 1949, 2,3.

15. *History of the General Staff*, 193.

16. Afterward, the cryptographic element in the General Political Directorate merged with the General Staff Cryptographic Bureau. Cde Nguyen Chu was appointed chief of the Cryptographic Section of the General Directorate of Supply. Cde Nguyen Dac Ho was placed in charge of military transportation cryptographic.

17. Consisting of comrades

Le Thanh Hai, political commissar of the General Staff Cryptographic Bureau; Cde Nguyen Trieu; Cde Le Van Chuong; Cde Le Van Bang

18. *History of the Vietnamese Communist Party*, first draft, 1920-1954. Hanoi: Truth Publishing House, 1984, vol. 1, 602.

19. In 1952, the Eastern Region Sub-intersector Cryptographic Section assigned three Comrades, Le Dzan, Nguyen Bao, and Nguyen Tan Nhon to go over and reinforce the Volunteer Army Cryptographic Section of East Cambodia.

Chapter Three

The Army Cryptographic Branch
Continues to Build and Develop in Every
Aspect, Serving Command Leadership, Developing
Guerrilla Warfare and Stepping Up Mobilization for
Progress into War of Movement (1951–1953)

CHANGE OF NAME TO ARMY ESSENTIAL MATTERS BRANCH;
CONTINUING TO BUILD AND DEVELOP,
EXPANDING ORGANIZATIONALLY;
RAISING CRYPTOGRAPHIC TECHNIQUE

After the Border defeat, the French colonialists fell deeper into a situation of perplexity and stalemate; therefore, they had to rely on the American imperialists to continue the war of aggression in Indochina.

Getting a shot in the arm by America, the French colonial gang consolidated and concentrated their pacification forces in an urgent pullback from the Tonkin delta to effect a policy of "taking war to breed war, using Vietnamese to beat Vietnamese" and prepare conditions for a counteroffensive to regain the strategic initiative. This was an all-out effort on the part of the French colonialists and the American interventionists.

In a nationwide atmosphere of elation after the Border victory, the second national congress of party representatives convened in February 1951. Resolutions of the congress brought out the mission of stepping up the resistance to achieve total victory. We had to build up larger armed forces –resolve to defeat every one of the enemy's warfare schemes. Implementing the resolutions of the congress, the Central Party promptly decided to reorganize the troops, striving to open a campaign aimed at sapping enemy strength, spreading guerrilla warfare, destroying the enemy's plan to consolidate his forces and pacify the [Red River] Delta, holding fast to our correct course of action in the strategic initiative in the Bac Bo [Tonkin] theater.

Executing the Central Party's decision, the forces of the Vietnamese National Army were massed to build additional main force divisions [dai doan], while at the same time opening a campaign to strike into the enemy's defensive perimeter in the midland and the Delta.

Faced with the requirements of the new mission, with respect to the building and operations of our armed forces, and implementing instructions from the General Staff and General Political Directorate, the army cryptographic branch diligently strengthened and built itself up, with respect to organization, raising the level of cryptographic technique for orderly construction and task methodology in order to meet the requirements of the new phase.

55

In February 1951, the eighth army-wide cryptographic conference was organized in Viet Bac. The conference examined the task of recent years and issued mission direction for the cryptographic task in coming years. Cde Hoang Van Thai, Chief of the General Staff, visited and spoke at the conference.

Based on real-world experience, the conference clearly determined the matter of professional instruction for army cryptographic organizations, namely, "we must latch on to a hierarchical branch system, instructions from above penetrating below, 'below' understanding 'above,' in order to fulfill the tasks swiftly, ensuring results." Wishing for successful outcome of this provision, "we must carry out the building and consolidation of the cryptographic organization to be sensible, scientific, unified, close," conforming to the organizational principles of the army; must build the specific responsibilities of the army cryptographic organization at the various levels, build a team spirit in the task between cryptographic organizations and people in command, communications organizations, and cryptographic organizations of the Party, government, and Public Security; build a system of administration for cryptographic cadre and personnel, raise the sense of responsibility, enthusiasm for the task, love of branch, love of skill in accomplishing good results in the specialty mission, etc. Resolutions of the eighth army-wide cryptographic conference also dealt with the matter of international obligation vis-a-vis Laos and Cambodia.

The conference also returned a proposal, and HQ made the decision to change the name of the army cryptographic branch to the army Essential Matters [co yeu] branch.

After the eighth army-wide cryptographic conference, cryptographic ["essential matters"] organizations at the various echelons in the army were unified and placed directly subordinate to the staff organizations at the various levels, under the direct control of the chief of staff. Tables of organization were gradually squared away and strength, dependent on the mission requirements of each unit. Responsibilities and mission of cryptographic organizations at the various echelons, from General Staff Cryptographic Bureau down to the cryptographic organizations at the level of regiment and provincial unit, and the mission of each specialized and responsible element--research, code compilation, message encrypting and decrypting, training and development--also in stages were built and fully worked out. The system for administering cadre and personnel, first being the system, procedures, regulations for selection, development and use, or for some time to perform the cryptographic task, also was promulgated and more closely implemented than before. Each relationship between the cryptographic organization and counterpart organizations was gradually squared away.

The General Staff Cryptographic Bureau was given an additional boost: Returning from study in China, Cde Hoang Van Dong resumed the duty of bureau chief, with Cde Nguyen Trieu the political assistant. Organizational structure of the bureau was changed to

- Campaign Cryptographic Section, with bureau deputy chief Nguyen Chanh Can as its Chief

- Encrypting-Decrypting Section, under Cde Hoang Manh Tuan

- Technique Research Section, under Cde Dinh Loan Thuyen

- Organization and Education Section, under Cde Le Van Bang

- Printing Team, with Cde Nguyen Tuan Nhan as team chief.

Cryptographic organizations in the Intersectors, divisions, regiments, provincial units, etc., also were matters of concern in the guidance for building. Cryptographic organizations distant from Central were unable to attend the eighth army-wide cryptographic conference, due to wartime conditions, so early in 1951 the General Staff assigned a cadre group under Cde Le Dinh Y, with Cdes Ho Si Dzi and Nguyen Tuan, to go down to Intersector 5 and Nam Bo to organize and guide the employment of cryptography in accordance with the new system and to see to it that the conference resolutions were fully grasped.

In May 1951, the Intersector 5 Cryptographic Section organized an intersector-wide cryptographic conference. It examined the research task situation, grasped the resolutions of the eighth army-wide cryptographic conference, and issued mission and cryptographic task measures for the Intersector.

In August 1951, Y's group reached Nam Bo. Because of task conditions, going and coming was difficult and dangerous, and it was urgent that [cryptographic capability] be kept together to serve command, but Nam Bo organized to comprehend thoroughly the import of the resolutions for the entire branch. Afterward, Cde Le Dinh Y received orders to remain in Nam Bo in charge of cryptographic technique training research for the Cryptographic Section of HQ, Nam Bo.

Also in 1951, the General Staff augmented Intersector 5 and Nam Bo with a number of experienced cryptographic cadre in order to increase the cryptographic organizations of their units.

So as to ensure that cryptographic organization at the various echelons was firmly secure, the General Political Directorate and the General Staff decided that Party membership would be a criterion in the selection of people for the cryptographic task, and instructed the commissars at the various echelons to take positive measures to educate and foster non-Party-member cryptographers, arranging for conditions to help them all train for entrance into the Party. In those instances in which aptitude for entry into the Party was lacking (or there were Party members lacking the conditions for the cryptographic task), they were to be transferred to a different task for which they were suited.

Along with the matter of strengthening organizationally, the General Political Directorate instructed the commissars at the various echelons and the unit commanders to attach special importance to leadership and ideological education for cryptographic cadre and personnel subordinate to them, to build responsibility vis-a-vis the specialty mission, raising enthusiasm in the task and clearly aiming at lengthy service in the army cryptographic branch.

With this sort of spirit, documents of the eighth army-wide cryptographic conference determined: "The cryptographic branch is an important branch in the service of the army, serving in the cause of revolution. . . .Viewed in this manner, we can endure long service for the cryptographic branch, possibly working five, ten, or twenty years and more in the cryptographic branch . . . . Viewed in this manner, we can straighten out our work, the conduct of people, in order to be worthy of performing the important task which has been entrusted to us by the association . . . . We must be of one mind from top down, so that each cryptographic person as one, each cadre and personnel as one--one and all--must be unified in outlook and ideology and also with respect to style of work, in order to serve the revolution--to serve the people: Unified in thought and viewpoint this way in order to create conditions to improve technique and speed along the development of the war."

The ideological viewpoint from above was regularly perceived by the cryptographic organizations at the various levels, educated in the stages of reeducation and troop reorganization and in internal life, so that it created a change with respect to perception and ideology for the cadre and personnel in the branch.

In the Intersectors and divisions, the task of education and ideological leadership for the cryptographic cadre and personnel was taken very seriously by the leadership and command comrades. In Intersector 5, Cde Nguyen Chanh, political commissar at Intersector HQ, got personally involved in educating and mobilizing the intersector cryptographic cadre and personnel and in determining the sense of responsibility vis-a-vis the mission, the profession, while at the same time deciding to augment the political cadre in order to build the intersector Cryptographic Section into a strong and stable unit. In Nam Bo, the cryptographic organizations also received serious consideration by the unit commands in their building-up. The Eastern Region Sector Committee assigned a sector current affairs committee member comrade to personally direct the political and professional reeducation for the cryptographic cadre and personnel.

In 1951, the army cryptographic branch also organized coordinated studies concerning the Party's revolutionary line and resistance line, its decision concerning the building of armed forces with the stages of professional reeducation, aimed at bringing into play the very essence of a people's army, a revolutionary army.

The cryptographic cadre study and training movement from 1951 spread far and wide, creating a seething momentum for the task. Cryptographic cadre and personnel regularly turned their thoughts to study and self-improvement, carrying out criticism and self-criticism, united to help one another in advancement. In this way, class position and sense of responsibility vis-a-vis the specialty mission of the cryptographic cadre and personnel clearly moved forward, most obviously in the matter of fixing responsibilities with respect to the specialty mission, building professionalism, and a sense of responsibility.

The Army Cryptographic School was officially established in accordance with High Command decision. On 14 May 1951, Comrade Nguyen Chi Thanh, head of the General

Political Directorate, personally assigned responsibility to a board of governors, comprising comrades Nguyen Dzuy Phe as director, Pham Tu Cap as deputy director, Vo Van Nhuong as the political assistant. The subject-matter-expert instructors were comrades Hoang Quyen, Vu Ngoc Hai, and Nguyen Mai Hanh. In order to meet the requirement for the number of cadre and personnel before the expansion of organization and technique for the cryptographic branch, the General Staff decided to open a class of instruction in the new technique, with the designator "C.40." The General Staff gave HQ, Intersector 4 the mission of organization, enrollment, and taking care of the logistics for the school. Through a selection process, considering all aspects of politics, ideology, education, and health, 250 students were selected from the provinces of Thanh Hoa, Nghe An, and Ha Tinh, from the Ground Forces Officers' School, and from the Intersector 4 Politico-Military School.

On 2 September 1951, opening day exercises were organized at the town hall of Hung Dao, Hung Nguyen district, Nghe An province. Attending were Cde Le Nam Thang, representative of the Sector Committee and HQ, Intersector 4, and Cde Nguyen Dinh Tung, responsible for political matters in Intersector 4.

The students were organized into four platoons. Squad cadre were taken from the students at the Ground Forces Officers' School and the Intersector 4 Politico-Military School. Platoon political personnel were taken from the students who had political assignments back in their units. In order to ensure secrecy and safety, the school regularly moved to training locations in the Nghe An province area of responsibility. Besides content that fostered the specialty profession, the curriculum of political study was also regarded with due care and attention, its content stressing "strive for self-improvement in ideology; train in the virtues of cryptographers who are party members" and prepare to accept and totally accomplish each mission entrusted by the Party. Students wrote resolute letters fixing responsibilities vis-a-vis the mission, as assigned after class sessions.

As for specialty content, compared with the study of technique KTA, training in technique KTB was much more complex, thus the acceptance of theory and the practice of encrypting and decrypting required the spending of time and labor, and mental power, before becoming proficient, especially in the basic subjects, such as cryptographic subtraction, memorization study, combined encryption-decryption, etc.

The movement to emulate teaching, study, cultural life and physical education – sports – to raise self-improvement and improve life--also boiled up regularly through mobilization.

The aspects of the task with respect to ensuring secrecy, guarding against traitors, and carrying out propaganda among the people were carried out seriously and strictly. The school organized sessions of labor to help the people and to participate in literary and artistic performances in the countryside. Executive committees,government, and people of the area strove to help out the school. Many mothers, such as Mother Dat in the village of Mau Lam, volunteered to give up their home for the students to have a place to study, mobilizing the women and children of the countryside to help out with sustenance and to take turns cooking for the students. Mother Dat's copper pot, used to cook for the students, was turned into an emotion-filled keepsake between the school and the people of the region, and continues to be retained by the school.

Building became an orderly routine – style of work, professional tasks, followed a regular direction, unified, taking into consideration the requirements for building and implementing in a positive manner the army's combat requirements. In order to build a basic foundation and get professional tasks into an orderly routine, in May 1951 the army cryptographic branch wrote up and the General Staff promulgated decisions with respect to

- The tasks of research, production, allocation for use, and maintenance of cryptographic materials.

- The tasks of encrypting and decrypting.

- The tasks of ensuring cryptographic security and the administration of secret messages.

- The meshing of tasks between the cryptographic organization and the commander.

- The important meshing of tasks with the communications and operations organizations.

In August 1951, the General Staff promulgated a reeducation document on staff professionalism, which included (Part B) relatively concrete regulations concerning task relations between the various echelons and the cryptographic organizations, principally in the matter of handling secret messages, requirements for education to promote a spirit of vigilance, security consciousness, and implementing specialty rules of conduct for cryptographic cadre and personnel. In Nam Bo, the Sector Committee and the HQ of the sectors issued regulations that cryptographic cadre and personnel of the various units would not be given leave to go into enemy rear areas. In Sector 4, the Intersector Committee and Intersector HQ instructed organizations having interconnected coordination with cryptographic organizations in the administration of cryptographic cadre and personnel to take the initiative in stopping and promptly settling negative occurrences that took place.

With a spirit of revolutionary vigilance, the army cryptographic branch had to take the initiative in diligently countering every one of the enemy's subversive plots and tricks. With respect to increasing the building of organization, the army cryptographic branch had to strive to raise the level of cryptographic technique in order to ensure secrecy for leadership and command under conditions in which the cryptographic liaison net was spread widely and deeply, the volume of cipher messages sent on the air had greatly increased, and the enemy was making every effort to collect cryptanalytic information in order to learn our cryptographic secrets.

The task of research to improve and enhance cryptographic technique in the army during this period had mobilized to produce a wave, arousing the ingenuity of all of the cadre and personnel in the branch. Many improved methods, enhancing the level of security of the code charts, were implemented, and there was research into the thorough development of ways of connecting up the Vietnamese language in the code charts, applying many methods of connecting up the language into one clever, creative method. There were places that concurrently used two methods of connecting up the language: In Nam Bo, the plain chart was filled out with clusters of syllables and phrase particles (called compound words) consistent with command vocabulary. There were places that used an auxiliary chart to contain compound words, places that arranged two plain elements in one cell and used many charts at a time, with general charts, special charts, charts for immediate use, and reserve charts. In form of compilation, the code charts were put in order, presenting a more scientific method of encrypting and decrypting that was favorable to speed and precision in conditions of mobile combat. With the above advances, the quantity of plain elements in the code charts increased noticeably, shortening the encrypting and decrypting of messages, speeding up message handling.

Together with the improvement of the plain chart [bang ro], the secret strip [bang mat] also was changed, in order to raise the capability of the cipher key to ensure secrecy. The secret strip was improved by many rows, many columns, or under the form of a set of strips, using an abridged set of letters to arrange cipher letters. The period of moving the short strip was irregular, using many strips together at one time. All of these methods of technique contributed to "frequency flattening," implementing nonrepeating [khong trung lap] substitution.

There was also concern for research – the compilation of cryptographic theory. The Nam Bo Cryptographic Section had compiled "The Theory of the Research Task and the Production of Cryptography." Although we had made strides, still we were immature with respect to other nations, because we had not inherited a legacy from the past, as had other nations, because we had no one skilled and professionally trained. Beforehand, the enemy had many centuries of experience – they had scholars in research, they had extremely clever compilers – our progress with respect to cryptography was not yet really remarkable. For these reasons,we could afford to be subjective. If we wanted the resistance to win unification and independence quickly to achieve a better outcome, if we wanted that success to be a solid building block in the foundation of world peace,we had to exert every effort to make our cryptography progress much further, so that we could catch up with advanced nations, etc.

Many forms of the KTA-type code chart differed from each other: Best known were the four-element, such as DzC-4, of the General Staff Cryptographic Bureau; the four-element of Nam Bo Cryptographic Section, that combined substitution with transposition; the three-element of Intersector 5; the five-element of Nam Bo; the six-element of the 320th Division. . . Generally speaking, the forms of the types of system were rich and dynamic. Super encryption was also researched and implemented. In Nam Bo's book, *The Theory of the Research Task and the Production of Cryptography*, superencryption was touched upon:

"... superencipherment is a method of using many methods of encrypting messages piled on top of each other. . . the method of encrypting messages by simple encipherment is inadequate to ensure the contents of the message --on the one hand, the groups of characters replaced in each system have limits; on the other, the volume of cryptograms sent in space and time accumulates. Therefore, the enemy has a rather rich quantity of materials to research and cryptanalyze our systems. . . ." "Superencipherment takes a lot of time and care on the part of cryptographers, but faced with requisites of the work, the cryptographer cannot flinch. . . ."

The cipher rules to implement superencipherment were applied as follows: "substitution of the Vietnamese language chart combined with advanced Playfair substitution," "chart substitution combined with transposition," "chart substitution combined with random key. . ." – these cipher rules were only beginning to be used in the larger units in order to ensure the secrecy of [message] contents [involving] strategy and campaigns.

Thus through improved processes, the code chart developed to a high level. Nevertheless, analyzed more deeply, it must be appreciated that, scientifically, a chart code has numerous weaknesses that cannot be eliminated in implementation--the rule of frequency, how the language is put together, the fixed form of the chart. Concurrent with the improvement and raising of the level of KTA, from April 1951, the General Staff Cryptographic Bureau pushed research to organize production of a type of cryptographic-- technique in which the encryption method relied upon a combination of code book ["cryptographic dictionary"] and mixed key, designated KTB.

This is a type of cryptographic technique of high grade, but with many complicated requirements with respect to production and design. The Technique Research element, newly placed under Cde Nguyen Dzung Hoa, overcame difficulties in searching out the frequency of command terminology in order to compile the code book. The demands of random key production had to be placed under a criterion of high randomness, during a time in which we had no calculating machines and had to use manual methods. Research into the compilation of codebooks and setting out a formula for cryptographic key production were technically researched with respect to content and method. The work had to be done in a very meticulous way; the labor expended was very time-intensive. After many months wrapped up in the work of the research team, code books and random key had been settled and production organized.

The comrade chief of the Cryptographic Bureau of the General Staff, together with the comrade chief of the Technique Research Section, carried out a cautious review and appraisal of the grade of security afforded by KTB, after which they jointly proposed that the General Staff permit its use in some large units. In December 1951 the General Staff decided on test-use of technique KTB on the mainline net between the High Command and the 308th Division.

The appearance of cryptographic technique KTB marked a new stage in the evolution of technique on the part of the army cryptographic branch. This was a type of technique that effected the substitution of plain and cipher in a largely random way.

At the beginning of 1952, the ninth army-wide cryptographic conference convened. Among the resolutions of the conference, a part spoke to the course of the mission in coming years, stressing "step up the training, reeducation and ideology to raise the quality of cryptographic cadre and party members; improve technique in accordance with the principles of secrecy, accuracy, and speed; strengthen organization; grasp firmly the guidance concerning the main theater of war; progress in unifying guidance on the entire theater of war," etc.

Comrade Nguyen Chi Thanh visited and addressed the conference. He said, "Comrades engaged in cryptographic work are anonymous warriors. But 'real talent needs no publicity.' You comrades must strive to give your very lives because of the revolutionary work of class and party, because of the work of resistance of the people. You comrades must strive to exchange revolutionary virtues and combat individualism in order to be worthy of the Party's trust."

By 1952 our army had formed six main force infantry divisions and one artillery-engineer [cong phao] division. Each intersector had two main force regiments. Nam Bo had four. The cryptographic organization in the divisions and regiments was boosted a notch. The use of KTB was expanded by the General Staff to the 304th, 312th, 316, and 320th divisions to enable direct contact with the High Command.

In April 1952, the General Staff Cryptographic Bureau opened the "Song Da" class to foster the new technique for a number of cadre in charge of Intersector and division cryptographic organizations in Bac Bo [Tonkin] to become the nucleus for training and organizing the use of the new technique in the Intersectors and divisions.

In June 1952, C.40, the class of training in the new technique, concluded. More than 200 student graduates were parceled out to the units to extend the use of the new technique.

Also in 1952, classes Quang Trung 1, Quang Trung 2, and Quang Trung 3 in turn were opened, training many additional cryptographic cadre and personnel to augment the units. The General Staff posted a number of cadre from the Cryptographic Bureau down to Intersector 5 and Nam Bo to help build the cryptographic organizations to the south solidly. A cadre group of three comrades – Luong Dzan, Nguyen Tat Giang, and Vu Dinh Son, with Cde Luong Dzan as chief – went down to Intersector 5 and Nam Bo to communicate the resolutions of the ninth nationwide cryptographic conference and to help the units organize political and professional studies for cryptographic cadre and personnel. After accomplishing this mission, Cde Luong Dzan was assigned as chief of the Western Area Subsector Cryptographic Section and Cdes Nguyen Tat Giang and Vu Dinh Son augmented the Nam Bo cryptographic section.

Apart from the increase by Central's cryptographic cadre, the Eastern Area Sector Committee issued instructions for the units to take note and obtain a number of Party members of worker and peasant stock and children of cadre to go and perform the cryptographic task at various levels. The Sector Committee also instructed the units to increase the educational task, the administration of cryptographic cadre and personnel, and to ensure that cryptographic organization was pure and solid.

Vis-a-vis the cryptographic organizations in the Bac Bo [Tonkin] theater, the General Staff Cryptographic Bureau attached special importance to increasing instructions for professionalism, organizing cadre groups to go down and provide on-the-spot assistance. In August 1952, Cdes Do Lac and Nguyen Nhien were sent to Intersector 3 and the 304th Division and the 320th Division. At the end of 1952, Cde Hoang Quyen was sent to the 325th Division in Intersector 4. Task groups helped the Intersector and division cryptographic sections organize political and professional training conferences and opened classes to develop new cryptographic cadre and personnel.

As a result, in the two years from 1951 through 1952, the cryptographic branch positively took the initiative to build and expand in every aspect – political ideology, organization, technique – and to direct the orderly building of the professional task, meeting the requirement to serve command leadership in conditions of very heavily armed forces and successive large campaigns.

SERVING COMMAND LEADERSHIP IN DEFEATING THE ENEMY'S URGENT PACIFICATION SCHEMES AND DECISIVE COUNTEROFFENSIVE

Right after the Border Campaign, at the end of 1950, the General Staff Cryptographic Bureau received orders from the Chief of Staff to prepare for the Trung Dzu Campaign (the Tran Hung Dao Campaign on the fields of Vinh Yen and Phuc Yen). The objective of the campaign aimed at continuing to wipe out enemy strength, expanding bases, developing guerrilla warfare to destroy enemy schemes to strengthen their forces and participation, and grasping the strategic initiative in the Bac Bo theater. The bureau laid out a plan to rectify and augment cadre and personnel, anticipate the organization of liaison nets, and arrange cryptographic material for the units participating in the campaign, consisting of the 308th and 312th divisions and the 36th, 88th, 102nd, 209th, and 141st regiments subordinate to the two divisions. Liaison nets to serve the Forward Area Supply Council, under the charge of the General Directorate of Supply, and the reconnaissance net, under the charge of the Intelligence Directorate's [Cuc Tinh bao] cryptographic, were also developed.

Drawing on the experience of the Border campaign, in this campaign we anticipated unfolding situations, organized to arrange for cryptonets to serve both units directly involved in the campaign and those on the sidelines – units in direct contact, skip-echelon, and cooperating in a relatively logical way.

The campaign began on 25 December 1950 and ended on 17 January 1951.

In the process of implementing the campaign, as it was opening the enemy launched a raid on Xuan Trach. The radio station of the campaign CP was unable to make contact with the 312th Division because the set was placed at the foot of Tam Dao mountain. Communications and cryptographic pooled their efforts to organize liaison for timely envelopment, moving the net via basic CPs on the other side in order to relay to the 312th Division. The campaign Forward Area Cryptographic Section, because it had taken precautions from the outset, had prepared and had cryptographic systems ready, so when this need arose to work this way, command was ensured throughout. Thus the 312th Division blocked and struck the enemy promptly, wiping out in its entirety the 24th North African battalion at Xuan Trach. This was also an experience in organization and implementation of the cryptographic task in a campaign: When the normal communications net between two units was interrupted, they could go via a third unit as intermediary in order to regain liaison, and, if one wanted to be able to do that in the plan that had been prepared, one had to have anticipated before the requirement came down.

However, in the conference summarizing phase one of the campaign, organized at the front on 3 January 1951, the campaign CP repeatedly called attention to the phenomenon of units writing reports but not checking the supervision and speeding up of transmission; therefore reporting messages were still continuing to sit at the station. The CP also observed: "One matter that must be paid attention to is the problem of maintaining secrecy when using radio and cryptography, when writing messages, copying messages, sending messages. . .cryptographic security is still a weakness."

In reviewing service to the campaign this time, the Cryptographic Bureau of the General Staff stressed, in addition to the value of planning and preparation, arranging cryptographic liaison nets, figuring out beforehand the circumstances calling for direct liaison, skip echelon, pooling and splitting up nets, moving nets, relaying via an intermediary--one very important requirement, namely, to track outgoing and incoming message flow; to settle each relationship between the cryptographic organization and the communications organization, operations, and command personnel tightly and promptly, regularly examining and digging out problems in the sphere of responsibility and mission of the cryptographic organization, along with interconnected problems of ensuring command requirements, and bringing them to the attention of command personnel for resolution.

The Trung Dzu campaign had just concluded when, on 20 March 1951, we opened the Hoang Hoa Tham campaign. Participating forces consisting of the regiments of the 308th and 312th divisions struck the enemy's defensive perimeter on Route 18, from Pha Lai to Uong Bi. After a series of indecisive battles, the campaign was concluded on 7 April 1951. The cryptographic forces of the campaign command and the divisions were diligent in service, but radio contact with the regiments still caused some hitches, and we missed a good opportunity to exterminate the enemy fleeing from Uong Bi and failed to get a timely order to the 36th Regiment to postpone its attack when information was received that the enemy had reinforced Mao Khe.

From 28 May to 20 June 1951, we opened the Quang Trung campaign in southern Intersector 3 (Ha Nam-Nam Dinh-Ninh Binh), the objective being to aid the guerrilla fighting in the Tonkin delta, with participation by the regiments subordinate to the 308th and 304th divisions. The division cryptographic sections and the regimental cryptographic teams taking part in the campaign accomplished their mission well, ensuring continuous encryption and decryption of command orders in the process of the campaign, especially the 304thDivision as a newly formed unit, plunging at once into a large campaign in the delta.

Through three consecutive campaigns, we struck the enemy at places where he had strong fortifications and had reinforcements of air and artillery and a high degree of mobility. Our divisions, for the most part, were newly brought together and built. We wiped out much of the enemy's strength (more than 10,000 men) but at the same time also extracted much experience in command leadership. The cryptographic organizations also came through many tests of their training.

The second Central Party Conference convened from 27 September to 5 October 1951 and estimated the situation and the enemy's schemes and clearly laid out our mission and activity guidelines for the coming period. The conference laid stress on the requirement for raising our quality in three types of armies, stepping up mobile warfare and spreading guerrilla warfare.

In October 1951, we opened the Ly Thuong Kiet campaign with the forces of the 312th Division striking the enemy at Nghia Lo (Northwest). The other divisions also prepared hard to open campaigns in many other theaters, sometimes taking turns training,sometimes fighting, with the form of quick-minded action.

Through a year of striving to strengthen his defensive positions, carry out pacification and increase his forces, the enemy plotted a counteroffensive to regain the strategic initiative. In November 1951 the enemy arrayed twenty battalions to strike and seize Hoa Binh, with the object of cutting our line of communication and supply, faced with attack and wiping out [by] our main force troops. The Main Military Committee decided to open the Hoa Binh Campaign with the forces of three divisions, the 308th, 312th, and 304th, striking the enemy on the main front; the 316th and 320th divisions would make a coordinated strike in the enemy's rear, in the Bac Bo [Tonkin] delta region. As for the cryptographic organizations taking part in the campaign, beyond their mission of ensuring service to command between the campaign CP and the divisions directly participating in the campaign, they would also have the constant mission of ensuring liaison between the campaign CP and the General Staff.

On the main Hoa Binh front, the cryptographic organization of the 308th Division under Cde Nguyen Than and the cryptographic organization of the 312th Division under Cde Nguyen Manh Mai served command, fighting in the principal direction of attack to break the enemy's Song Da line.

The cryptographic organization of the 312th Division worked closely with communications on the route of advance in order to ensure timely encryption and decryption of operational orders in a mobile battle of ambush that wiped out the enemy's 1st Parachute Battalion at Ninh Mit, southwest of Ba Vi mountain, after which it continued good service to the division command in striking the enemy throughout the process of the campaign.

The cryptographic organization of the 308th Division served in handling command orders accurately and speedily in the assaults of the 88th Regiment on the entrenched fortifications at Tu Vu, afterward serving the command of the 36th Regiment striking the enemy in Hoa Binh township.

The cryptographic organization of the 304th Division, under Cde Ngo Duc Tri, served command in combat on routes 6 and 21, speedily encrypting and decrypting orders deploying the 66th Regiment from the Hoa Binh front back to strike the enemy at Thuong Tin, Ha Dong, opening an additional direction of pressure on the enemy to the south of Hanoi.

In the front behind the enemy, cryptographic of the 316th and 320th Divisions entered the delta region of Intersector 3 and the midland to serve in attacking the enemy in cooperation with the Hoa Binh front. In conditions of activity in the region under temporary occupation, we had to continuously cope with encirclement, raids, and the enemy's lethal weapons, working in underground shelters that lacked light, then off into mobile operations, but the cryptographic cadre and personnel overcame difficulties and continued to achieve the requirements of command in hitting the enemy on the fields of Bac Ninh, Phuc Yen, Route 5, Ninh Binh, Nam Dinh, and Phu Ly, ensuring cryptographic security. Unit cryptographic cadre and personnel encrypted and decrypted thousands of High Command messages quickly and accurately. Especially noteworthy was that almost every HQ message providing timely reporting of enemy preparation to raid and develop our strength--principally the times the enemy had forces prepared to encircle us, intending to wipe out our main force [units]around Nam Dinh and Ha Nam--which cryptographic above and below speedily encrypted and decrypted, enabled the army and people of the region to take the initiative in facing and thwarting the enemy's encirclement schemes. Service to cooperative operations in the enemy's rear between the 320th and 304th Divisions also was organized and well taken care of.

At the end of the first lunar month of 1952, the enemy pulled out of Hoa Binh, ending the Hoa Binh campaign: we wiped out 22,000 of the enemy, forced surrender, forced the evacuation of many of the enemy's posts and entrenched fortifications, and expanded many guerrilla bases in the region of the enemy's rear. The enemy's scheme of pacifying the Bac Bo delta and a counter offensive to regain the initiative had been defeated.

The army's cryptographic organizations at all levels had ensured the accomplishment of service to command, participating in a common victory. The Cryptographic Bureau of the General Staff had accomplished its mission of command service to the High Command and the General Staff vis-a-vis the intersectors and divisions, and many times went directly down to the regiments and battalions for timely service in striking the enemy, dashing out, avoiding losses. Through the realities of the campaign, both in the main direction and the secondary direction, on a broad area of responsibility, we had still ensured cryptographic security, while, at the same time had also diligently decrypted messages of the enemy, discovering and meeting in timely fashion many tricky encirclement actions and pullbacks by the enemy, and serving operational guidance. Otherwise, the Bureau continued to perform well the function of guiding the cryptographic organizations at the various levels, army-wide, in the realization of the specialty mission of each echelon. With the accomplishments it had achieved, the General Staff Cryptographic Bureau was awarded the Order of Military Merit [huan chuong Chien cong], second class.

****

At the beginning of September 1952, the Politburo decided to open the Northwest campaign, aimed at sapping enemy strength and liberating part of the Northwest. Forces participating in the campaign consisted of the 316th, 308th, 312th, and 351st divisions and the 148th Regiment. The 304th and 320th divisions would carry out a coordinated strike against the enemy in Intersector 3.

In the opening battle of the campaign on 14 October, the cryptographic teams of the 174th Regiment (316th Division) and 141st Regiment (312th Division) encrypted and decrypted the orders of the campaign command post, commanding the units to wipe out the Ca Vinh and Sa Luong posts, etc.

Afterward, the cryptographic organization of the 308th Division encrypted, decrypted and passed on the orders of the campaign CP and of 308th Division HQ commanding the 102nd and 88th regiments to strike the enemy in the Nghia Lo Village and the 36th Regiment to wipe out the Cua Nhu post. With the Nghia Lo Subsector wiped out, the cryptographic organization of the 312th Division served the command of pursuing troops over four successive days and nights.

In the second phase of the campaign, the cryptographic team of the 148th Regiment and that of the 165thRegiment (312th Division) encrypted and decrypted [message traffic] to serve the command of troops thrust deeply in envelopment into the enemy's rear at Son La, Lai Chau, Tuan Giao, Dien Bien [Phu], etc.

On 10 December 1952, the Northwest Campaign was over, the cryptographic teams of the units participating throughout the campaign having closely coordinated to ensure a favorable outcome of the mission for command of the various wings of the army, and coordinated command between the forces in the front, those in the envelopment and in the battle in the enemy's rear.

The period 1951-1952 was one in which the army cryptographic branch built and expanded in all aspects, having established stages of change and progress all over, with respect to political matters, ideology, organization and professional technique. Especially, the branch had created ranks of cadre and personnel with solid political qualities, a basic level of specialty technique, and had researched and compiled a new type of technique with a high degree of security. Thus the army cryptographic branch strove upward to sufficient capability to meet the requirements of command in combat in a string of continuous campaigns opening up on a large scale, coordinating in many directions, and between theaters across the entire nation.

In particular, the matter of increasing the cryptographic branch's service to direction and command of theaters behind enemy lines encountered obstacles defeating the enemy's pacification schemes – occupying heavily populated areas with much property, carrying out his policy of "taking war to breed war, using Vietnamese people to beat Vietnamese people," making a situation in which the theater behind enemy lines changed to be valuable to us; moreover, the enemy's posts there were difficult and on the defensive.

Along with the above accomplishments, one must speak of the dazzling examples of the doctrine of revolutionary heroism on the part of cryptographic forces at the various levels. These comrades not only stood hardship and sacrifice, overcoming each difficulty to accomplish the mission, but still displayed a firm nature when falling into enemy hands, protecting the security of codes, maintaining professional secrecy, worthy of the esteem of the Party and the army. Here are some representative examples:

In June 1951, Cde Doan Thi Chat, a cryppie of Vinh Tra district, while performing her mission, had her secret underground shelter discovered during an enemy sweep. Cde Chat and a teammate organized to fight, concealed the cryptosystems and struck back at the enemy until the last breath, protecting the security of the technical documents. Also in June 1951, Cde Le Hoang Ninh, cryppie of Tam Binh district, Vinh Tra province,was discovered by the enemy in his secret underground shelter and called on to surrender. After concealing the cryptosystems, he waited for the enemy to get close enough to use a grenade to wipe out the ones surrounding the underground shelter. Knowing that he could not escape capture by the enemy, the comrade used a grenade at the end, exterminating the enemy and heroically sacrificing himself. Cdes Doan Thi Chat and Le Hoang Ninh were both awarded the Order of Military Merit, first-class. The cryptographic team of the 101st Regiment, 325th Division, performing its mission in the enemy area of [Quang] Binh-[Quang] Tri-[Thua] Thien, was ambushed by the enemy, captured, and held prisoner at Hue. They were savagely tortured, but the comrades held fast to the pride of being revolutionary warriors, right up until they drew their last breath.

Cryptographic comrades performing their mission in the Lower Laos-Highlands theater all fell ill with malaria. At times when command combat requirements were urgent, they alternated between being in bed and encrypting and decrypting messages, providing prompt contact for their unit. The cryptographic comrades assigned in the Extreme South [of Intersector 5] theater had to endure many adversities--there were long periods in which they were only supplied with 200 grams of rice and a mug of water a day. Many of these comrades were short of rice, lacked salt, lacked water, and they became ill, unable to see.

Faced with this difficult, arduous situation, the upper echelons concerned themselves with the cryptographic cadre and personnel, that they rotated into the free area to recruit their health and engage in professional study.

With accomplishments and also with lessons learned through experience in the years 1951–1952, there had been created conditions for the branch to move upward in accomplishing its mission during the strategic counteroffensive of 1953–1954.

Lieutenant General Hoang Van Thai, Central Party Committeeman and Deputy Minister of National Defense, speaking at the Conference to summarize the Army Cryptographic Task in the Service of the General Strategic Office of Spring 1975

General Le Trong Tan, PAVN Chief of the General Staff, at the celebration of the fortieth anniversary of the establishment of the Army cryptographic branch (September 1985)

Lieutenant General Doan Khue, Chief of the General Staff, visits the work spaces of the General Staff Crypto Directorate (1987)

The encrypting-decrypting element of the General Staff Forward Crypto Section during the TRAN HUNG DAO Campaign (December 1950)

A cryptographic training class in Viet Bac during the resistance against France

Message from the High Command directing that tactical operations be enciphered and deciphered secretly, swiftly, and accurately

Autograph of Cde Hoang Van Thai to the Army crypto branch

Chapter Four

The Army Cryptographic Branch
in the Winter-Spring Strategic Offensive of
1953–1954 and the Dien Bien Phu Campaign (1953–1954)

As we entered the winter-spring of 1953-1954, the resistance of our people to the French colonialist aggressors had entered its eighth year. The developing situation of the war was useful to us. The more we struck, the more victories; the more we struck, the stronger we were. As for the enemy, the more the fighting dragged on, the more he lapsed into defensiveness and embarrassment.

In January 1953 the fourth congress of the Central Party's Executive Committee convened. In Uncle's report, read at the meeting, Uncle analyzed and assessed the situation, and the stubborn nature of the enemy. Uncle showed clearly that "At the beginning of 1952 they lost big in the Hoa Binh campaign. At the end of 1952, they lost big in the Northwest campaign." "The more the enemy loses, the more brutal he becomes," thus "from now on, the war between ourselves and the enemy will become tougher and more complex." In order to move from resistance to total victory, Uncle stressed two principal problems: resistance leadership and military policy; thoroughly mobilizing the masses, reducing land rent and moving to land reform.

The resolutions of the congress also brought up clearly that "Our army must strike the enemy where he is weakest; at the same time we must be heavily engaged behind the enemy." "Whether in the mountains or in the delta, our army must certainly strike the enemy's forces and his ever-strengthening fortifications."

Implementing the line of the resolutions of the fourth Central Party congress, along with the Party's mobilization of the masses to implement the land policy, the great undertaking of building and raising the quality of our armed forces received special attention.

In March 1953, the Main Military Committee passed a resolution concerning getting the troops reeducated politically, the aim being to "Raise the level of class consciousness of the troops another notch, making organizations pure and solid, in order to heighten the combat capability of the troops, in order that the troops will become a larger, stronger force, determined to aid in implementing the land policy of the Party and government."

In the political reeducation classes for middle and upper level cadre, Uncle also taught "The aim of reeducating the army is to make our army into a revolutionary people's army determined on victory."

Through study of the [Party] line and the policy of mobilizing the masses to implement the Party's program of land reform, and through reeducating the army, along with the entire army, the strengthening of ideology and the organization of the army cryptographic branch was increased manyfold. Cryptographic cadre and personnel determined that the proletarian class position was clearly distinguished by the line between worker and exploiter; from that they increased their patriotism, felt hatred for feudal imperialists, strengthened their love of internal unity, and added to their zeal to strive on to accomplish their speciality mission.

After the stage of reeducating the army politically, the army cryptographic branch was strengthened with respect to organization from top to bottom, making its organization pure and strong. Many cadre Party members came from a working class or farming background, and through tests and training had been selected to augment into the army cryptographic branch. Organizational tasking and cadre in the army cryptographic branch became the objects of concern for the upper echelons and were given closer leadership and guidance.

In March 1953, more than fifty cadre from platoon to battalion level and nearly 200 cadre and party members from the regions were selected to attend the Army Cryptographic School in order to prepare to carry out the cryptographic task. At this time the Army Cryptographic School had the designator C65, with Cde Le Thanh Hai the political commissar and director.

Faced with the urgent requirement for service in the 1953–1954 Winter-Spring Strategic Offensive with the above number of students, the school immediately organized a short-course class in technique and professional knowledge, in order to swiftly augment the units participating directly in the Winter-Spring Offensive; the remaining number of students continued their study according to the syllabus and basic plan until peace was restored. By the end of 1953-early 1954, a number of students were designated to participate in mobilizing the masses to implement rent reduction and land reform.

Also in March 1953, the General Staff appointed Cdes Nguyen Dzuy Phe and Hoang Quyen to go inspect and assist Intersector 5 Cryptographic. HQ, Intersector 5 also selected a number of comrades from units to enter the cryptographic branch, such as Cdes Van Kien and Nguyen Thu. In Nam Bo, the cryptographic organizations were also strengthened and cadre added. At this time the Nam Bo HQ organizations had moved to the Western Area. The Western Area Subsector Cryptographic Section had been merged with the Nam Bo HQ Cryptographic Section, Cde Luong Dzan in charge.

In September 1953, our army commenced reeducation in military matters. As the Main Military Committee clearly indicated, the goal of military reeducation was that "we must again improve and train to be good at tactical technique. We have advanced a step with respect to politics and ideology; now we must advance in tactics and technique so as to have the combat capability to move up to the new stage with the army."[1] Bringing into play the results of the military and political correction, the cryptographic organizations throughout the army carried out emulation in study, raising the level of usage of the various types of cryptographic systems. Research, development, and production of cryptographic systems was speeded up. The Cryptographic Bureau of the General Staff had improved and raised the level of technique of the various types of systems, system DzC4, assault systems and spell-chart [BA-RA-XO] systems. The Cryptographic Section of Intersector 4 had researched and come up with the Chien Thang [VICTORY] type systems and 4-element Hoa Binh [PEACE]; Intersector 5 had invented a 3-element system; intelligence [tinh bao] cryptographers had invented the Doc Lap [INDEPENDENCE] system; the 320th Division had invented the "6/320" system, etc.

The improved systems just mentioned came about through experience drawn from real-life combat service, so they displayed abundant contents, were light and compact in composition, handy for enciphering and deciphering under conditions of combat in the field, mobile operations, missions behind enemy lines, in jungle and mountain, and other, different areas of operation. In order to take the initiative in the process of serving command in battle, the main force divisions [dai doan] of HQ were set up for concurrent use of both KTA and KTB in order to meet the urgent requirements of the new mission.

From the end of September 1953, the General Staff Cryptographic Bureau implemented an assignment plan to augment the cadre and personnel in the units, while simultaneously researching the arrangement of a number of cryptographic nets in the command system of the campaign and cooperating theaters. The scope of the nets was quite large and complex, comprising nets for skip-echelon, direct contact, and joint liaison, spread out over all of Indochina in these directions:

- Cryptographic organizations of the 316th and 308th divisions would move up to serve the command of units on the main Northwest theater.

- Cryptographic of the 101st Regiment of the 325th Division, and the Cryptographic Team of the 66th Regiment of the 304th Division, along with the units, would move across to Central and Lower Laos to combine operations with the Pathet Lao Liberation Army and the Cambodian Liberation Army, serving liaison with the General Staff and HQ, Intersector 4.

- Cryptographic of HQ, Intersector 5, together with the cryptographic teams of the 108th and 803rd regiments would serve the Highlands [Tay Nguyen] front.

- The cryptographic organizations of the 312th division and the engineer-artillery [cong phao] division, and the cryptographic teams of the 9th and 57th regiments (304th division) would ensure the command responsibility of the operation and divisions, with a view to distracting enemy forces in various directions.

- Cryptographic of the 320th division and the cryptographic teams of the 42nd, 46th, 50th, 238th,and 246th main force regiments would serve the fighting in the enemy's rear in Intersector III.

- The Eastern Area Sub Sector cryptographic organization of Nam Bo Western Area and the Saigon-Cholon Special Sector would ensure combat command in the Nam Bo theater, in coordination with the main theater.

In mid-November 1953, in accordance with strategic direction chosen by the Main Military Party Committee and the High Command, the cryptographic branch made sufficient preparation to ensure continuous liaison from HQ to the various directions and to efficiently serve the appointed responsibilities HQ had given the forces in the various directions.

On 26 November 1953, the General Staff Cryptographic Bureau appointed a task team under Cde Nguyen Cong Khuong to serve Cde Deputy Chief of the General Staff Hoang Van Thai, going to Northwest to personally direct operations in the Lai Chau Campaign.

Ferreting out the fact that our regular forces were appearing in the direction of Northwest, on 20 November the French dropped paratroopers on Dien Bien Phu, intending to help their army at Lai Chau. Discovering that the enemy was preparing to pull out of Lai Chau, Cde Hoang Van Thai sent a Flash message to the 316th Division: "On 6 December 1953, [French Gen. Rene] Cogny issued orders for the French army to pull out of Lai Chau. One element of the enemy army will be transported by air. Those remaining will withdraw by road, and must be completely out by 12 December. The division is ordered to quickly have an element follow Route 41 and strike into the town, while a large element goes to Tuan Giao by the short cut through the Pa Thong pass and cuts the Lai Chau-Dien Bien Phu road in order to wipe out the withdrawing army."

At the same time of this message, HQ also ordered the 308th Division to surround the enemy at Dien Bien Phu and block the road to prevent their running over to Laos.

Executing the above orders, on 10 December 1953 the 316th Division quickly assaulted the enemy at Lai Chau. After a few days of fighting, the 316th Division defeated twenty-four enemy companies, liberating the Lai Chau area.

The Lai Chau victory had great significance, for it was the victory that opened the Winter-Spring Strategic Offensive of 1953–1954. The cryptographic teams of the 174th and 198th regiments had stuck close to the units that staged a forced march along a shortcut to intercept the enemy pulling back from Lai Chau to Dien Bien Phu. Throughout many days and nights through the jungle – crossing mountains, crossing rivers, enduring hunger, enduring cold, participating in the operation – they ensured that messages would get out.

On the Central Laotian front, the cryptonet between the High Command and the campaign CP with the 325th Division, the 101st Regiment (325th Division), and the 66th Regiment (304th Division) on the march into action was maintained tightly. Divisional liaison coordinated operations between our army and the Pathet Lao Liberation Army and the [Cambodian] Isarrac Liberation Army attacking the enemy and liberating large parts of the Central and Lower Laos sectors, and northeast Cambodia, linking northeastern Cambodia bases with the liberated regions of Central and Lower Laos.

On the Intersector 5 axis, on 20 January 1954 the enemy army mobilized six mobile groups, with navy and air support, to attack toward Phu Yen, intending to attack and occupy the entirety of the Intersector 5 free area. The High Command and the Intersector

HQ--through the medium of the cryptographic liaison net--issued instructions in a continuous, timely manner to realize a plan of attack on the enemy's weak spots in the Highlands, to wipe out the enemy's entrenched fortifications and outposts, liberating all of Kontum City and the northwest Highlands and protecting the Phu Yen free area. After serving command in striking the enemy in the Highlands, the Intersector cryptographic organization continued on to serve the units striking the enemy in coordination with the Dien Bien Phu front.

On the Upper Laos front, in accordance with orders from HQ, the 308th Division made a forced march to attack and shatter the enemy's defensive perimeter in the Nam Hu river basin as a diversion and to isolate the Dien Bien Phu entrenched fortification. The division's cryptographic organization ensured that encrypting and decrypting were adequate, timely, and accurate for command communications throughout the process of the attack and the pursuit of the enemy.

On the other fronts, from the northern delta to Nam Bo, they stepped up attacks to sap the enemy's strength, to open guerrilla bases, and to coordinate with the main theater.

In the campaign of strategic assault in the winter-spring of 1953–1954, the Politburo had anticipated having much capability to bring about a great storm of a battle in the northwest. The thinking of the Main Military Committee was to fix the enemy at Dien Bien Phu and possibly resolve to fight there. As for the enemy, they decided to turn Dien Bien Phu into an "impregnable" fortress, preparatory to wiping out our main force troops.

Dien Bien Phu was to become the most decisive measure of strength between ourselves and the enemy in the Winter-Spring 1953–1954 Campaign.

On 6 December 1953, the Politburo of the Central Party decided to open the Dien Bien Phu campaign and as communicated via the combat operations plan of the Main Military Committee, the Politburo decided to establish a Party Committee and Dien Bien Phu Front CP, with Cde Vo Nguyen Giap, member of the Politburo of the Central Party and head of the High Command, as secretary of the Party Committee and Commander-in-Chief of the front.

In order to fulfill the mission of service to the campaign, the Campaign Cryptographic Section, under Cde Nguyen Chanh Can, issued guidance and instructions to the cryptographic organizations of the divisions and units participating in the campaign, directly expounding all facets of the assignment in preparing for the campaign. With the coordinating fronts, it was also necessary to monitor and carefully guide the [cryptographic] techniques in order to ensure continuous liaison in each circumstance.

While preparing to carry out the campaign, the cryptographic organization also had to ensure accurate encrypting and decrypting of each piece of news, concerning the activities of the enemy, concerning the preparation of the battlefield, concerning the Party and political tasks on the Dien Bien Phu front and in the theaters, serving the echelons figuring out the nasty schemes of the enemy, and examining, supervising, and speeding up each activity in preparation for the campaign.

By the end of December 1953, the organizational framework of cryptography devoted to serving the Dien Bien Phu campaign took shape:

- Cryptographic Section of the Dien Bien Phu Campaign Command Post.

- Cryptographic Section of the 308th Division with the cryptographic teams of the directly subordinate regiments.

- Cryptographic Section of the 312th Division with the cryptographic teams of the directly subordinate regiments.

- Cryptographic Section of the 316th Division with the cryptographic teams of the directly subordinate regiments.

- Cryptographic Section of the 351st [Engineer-Artillery] Division with the cryptographic teams of the directly subordinate regiments.

- Cryptographic of the provincial units and regional battalions subordinate to the Northwest Sector.

- Cryptographic Section of the General Supply Directorate with the cryptographic teams directly subordinate at the military relay stations.

- Cryptographic team of Section 2 [i.e., military intelligence – G2] with the cryptographic teams of directly subordinate reconnaissance [units].

- Cryptographic teams in Intersectors III, IV, and Viet Bac to ensure campaign command, transportation, and supply.

By 25 January 1954 each preparatory task for the campaign had been accomplished: our army and people were ready to attack the enemy at Dien Bien Phu. But right at this point, after having carefully considered all aspects, the Party Committee and Campaign CP decided to postpone the opening gun. This decision having been developed and implemented, cryptographic sent the Flash message from the Campaign CP: "It has been determined to change the operational approach [phuong cham, lit., "line"] from 'fast strike, fast resolution' to 'steady strike, steady advance,' although many difficulties must be surmounted," at the same time transmitting all sorts of message orders and plans of the CP, and the staff, political, and rear services organizations to all of the campaign units to hold their fire.

With the determination to change having extremely serious significance for the Dien Bien Phu campaign, the instructional contents from the campaign CP to the units via secret message had to be exactly accurate and absolutely secret, for the time element was very pressing: if there were only a small error in a sentence, or arrival a minute late – let alone an hour – it could have great influence on the outcome of the campaign.

The cryptographic organizations, from the Cryptographic Section at the Campaign CP to regimental cryptographic, all searched for methods of organizing the work of encrypting and decrypting, transmitting messages as fast as possible, allotting work sensibly, holding tightly to the liaison net, closely pooling efforts with the radio station to track messages, and for ensuring that secret messages were sent accurately and promptly to satisfy command requirements.

Together with serving the campaign CP's command over directly subordinate units participating in the campaign, the matter of maintaining cryptographic liaison with the leadership organizations of the Party and the High Command, also as with the cooperating theaters, had especially serious importance. Daily, through the Cryptographic Bureau of the General Staff in the rear, the Campaign Cryptographic Section regularly encrypted and decrypted the instructional views and communiques of the Central Party Politburo vis-a-vis the process of command and instruction for the campaign, at the same time sending messages from the Comrade Commander-in-Chief in the Dien Bien Phu campaign to the theaters to step up their cooperative actions and receiving news of victories by our armies and people in various places.

13 March 1954 was the day set for opening fire and attacking the enemy, opening the campaign. The system of cryptographic technique from campaign CP to the divisions and regiments had been augmented, arranged, and fully worked out from the time of the order postponing the attack on 25 January 1954, and combat service preparations had been ensured.* The Cryptographic Section of the 312th Division, the Cryptographic Team of the 141st Regiment, and the Cryptographic Team of the 209th Regiment served the command of the troops attacking the enemy at the Him Lam entrenched fortification [French strongpoint "Beatrice"]. Then, on 14 March the Cryptographic Section of the 308th Division and the Cryptographic Team of the 88th Regiment, together with the Cryptographic Section of the 312th Division, the Cryptographic Team of the 165th Regiment, and the Cryptographic Section of the 351st Engineer-Artillery Division ensured service to the command of the troops attacking and wiping out the enemy at the Independence Hill entrenched fortification [Fr. strongpoint "Gabrielle"]. After five days of fighting, we totally wiped out two of the enemy's first class and strongest entrenched fortifications. "Victory in the initial assault foreshadowed total victory for the campaign. It proved the line 'steady strike, steady advance' was totally correct. It marked the growth of our infantry, communications, artillery, and antiaircraft in combined branch combat."

At 1730 on 30 March 1954, we began to fire on the crests of the hills on the east. Cryptographic of the 312th Division (with the 141st and 209th regiments) and the 316th

---

* The history of the PAVN signal corps expands on communications and communications security arrangements for the Dien Bien Phu campaign. Front CP to divisions is said to have used OPCODE via the World War II vintage U.S. AN/GRC-9 and SCR 694; regiment to battalion used jargon code via the U.S. BC 1000 (SCR 300) "walkie-talkie" backpack, and battalion to company used the PRC 702 "in the clear." Artillery regiments issued fire orders in the clear. In addition to radio, single-wire field telephone arrangements were extensive. Signal panels (blue-white) and semaphore flags (red/white) were used and, at the squad level, bugles and whistles. (*History of the Communications-Liaison Troops*. Hanoi: Communications-Liaison HQ, 1985. Vol. I, 303-304.)-- Tr./Ed.

Division (with the 98th Regiment) served the command of the troops hitting hilltops C, E, and hill D ["Claudine," "Elaine," "Dominique"]. In this stage of the assault, the most decisive strike took place on hills A1 [part of "Anne Marie"] and C1, ending with us and the enemy each holding half of a hilltop. The assault on the eastern sector temporarily halted on 5 April 1954. In this assault stage, we had seized a large part of the important high points, and wiped out 2,500 of the enemy.

After opening the second assault stage, we continued to encircle and cut off and smash the enemy positions.

As of 1 May 1954, we opened stage three of the assault, into the heart of the Dien Bien Phu entrenched fortification, aiming to attack and fully occupy the high points on the eastern side, restricting the sphere of the enemy on the western side, wiping out the heart of the entire entrenched fortification in a general assault.

The cryptographic organizations of the units participating in the fighting served tactical operations command:

The 98th Regiment (316th Division) wiped out the enemy on hill C1; the 209th Regiment (312th Division) eliminated fortifications 505A and 505 at the foot of the hilltops on the eastern side, and the left bank of the Nam Rom river; the 88th Regiment (308th Division) struck the enemy at position 311A on the western side; the 57th Regiment (304th Division) struck into Sector C northeast of the southern subsector; the 36thRegiment (308th Division) wiped out position 331B, etc.

By the night of 6 May, orders for the general assault were speedily transmitted to the commanders of the units. Our army was divided up to make many points of attack on the positions: hill A1, hill C2, position 506 north of the Muong Thanh bridge, position 310 on the western side, restricting the enemy's holdings. The next day, 7 May, our army wiped out the enemy in the positions near the Muong Thanh Bridge and the left bank of the Nam Rom River, striking into the center, advancing straight to the enemy CP.

The afternoon of 7 May 1954, a report from the commander of the 312th Division was passed to the Campaign HQ: "All enemy forces in the central sector have surrendered. [General] de Castries and all of his staff are taken."

That night the Political Commissar of the 304th Division reported to Campaign HQ: "The 304th Division has taken alive the entire enemy headquarters at Hong Cum [Fr., "Isabelle"] on the run; we have Colonel Lalande."

Also that night, from the CP of the Dien Bien Phu campaign, the campaign cryptographic cadre and personnel encrypted a message from the Commander-in-Chief of the campaign reporting the glad news of the victory to the Politburo and Uncle Ho, and to the theaters of war: "At 1730 hours 7 May we wiped out the entire concentration at the Dien Bien Phu entrenched fortification. The Dien Bien Phu campaign is victorious." Just a half hour afterward, the report of our army's great victory at Dien Bien Phu was sent from the General, Commander-in-Chief of the front, to the Politburo and the revered and beloved Uncle Ho.

The 1953–1954 Winter-Spring Strategic Offensive of our army and people was concluded by the glorious Dien Bien Phu victory. The cryptographic branch of the army had fulfilled its responsibility to serve the various levels of command, from the High Command to the units participating directly in the campaign and the coordinating theaters of war, doing their appropriate bit in making this a historic feat for our race.

Through organization and instruction in cryptographic technique to serve command in the Winter-Spring 1953–1954 and the historic Dien Bien Phu campaign, the army cryptographic branch gleaned valuable mission experience in campaign cryptography.

The strategic offensive opened in many directions, combining many forces, with many segments, over the largest sphere experienced to date, with a liaison net expanding very widely, while the cryptographic organizations army-wide were using differing cryptographic techniques from each other. At the High Command and the main force divisions participating directly in the Dien Bien Phu campaign, they had started off using KTB concurrently with the use of KTA. In the Intersectors, the units only used KTA, but also had many different forms, the Forward Area Cryptographic of HQ using all types of single encipherment spell-charts [BA-RA-SO] and spell-charts superenciphered by random [loan] key. Through requests for service and technical conditions, therefore, the job of arranging the cryptonet, issuing instructions, and organizing the use of cryptographic technique was quite complex and difficult.

Because of thoughtful research and anticipation concerning the developing situation in the implementation of the campaign, and thoughtful preparation for every aspect of technique, and organization in the use of cryptographic technique to have one tight, sensible method in each of the various directions, then, from the outset, we ensured that we could fully grasp the situation, take the initiative, and favorably resolve situations arising "out of the blue," requiring immediate attention, in order to meet in every respect the requirements of command in the principal theater and even in the cooperating theaters.

In realizing the use of cryptographic technique to serve the campaign, the cryptographic organizations clearly perceived that the requirements of the command task were increasing and becoming more urgent with every passing day, and that these requirements were expanding rapidly. Striving to satisfy the requirements of the command task is to strive without cease, and the army cryptographic branch wanted to fully satisfy command requirements, which meant raising and improving every aspect of technique. For the internal structure of chart systems, [finding] a method of arranging the plain-cipher parts in a way that was sensible and most favorable for the task of enciphering and deciphering; the spell-chart format, while ample, when used for enciphering and deciphering took long to find the plain units, thus limiting efficiency, inconsistent with conditions of mobility and combat. The technical shortcomings were overcome, however, by the training of the enciphering and deciphering people but continued to influence the flow of messages and their timeliness. In the Dien Bien Phu campaign, the 36th, 102nd, and 165th regiments also had situations of backlogged messages to encipher that were not finished. The 102nd Regiment (308th Division), on the march eight kilometers from Dien Bien Phu, had orders to turn around and take a different road, but because the message arrived late, the unit continued its advance, encountered the enemy, and were forced to fight. The 308th Division was on the march to Son La,when HQ sent a message to division HQ and the units to halt and await orders, but because that message arrived late, the troops continued on some twenty km before they received the message and had to turn around. (In fact, the delay in receipt is still unresolved, for different reasons.) In circumstances of our pursuing the enemy or making a very fast movement, the time of stopover was only quite short, but the requirement for transmitting orders and instructions in these conditions was very urgent, and, if the message volume was large, cadre and personnel encountered not a few difficulties, but cryptography prevailed, despite all obstacles.

****

After the historic Dien Bien Phu victory, the Geneva accords were signed. The responsibility for serving the task of leadership and direction in the implementation of the accords and effecting the cease fire, regrouping the armies, and taking over the liberated region was established as being very urgent and by no means uncomplicated.

In July 1954, cryptographic cadre and personnel of our army in Laos, Cambodia, Nam Bo, and Intersector 5, together with Southern troops and compatriots, regrouped in the North. Twenty-five army cryptographic cadre and personnel were selected to remain in the South, and a team of two cryptographic comrades was arranged to stay in Cambodia to continue to serve revolutionary responsibilities in the new situation.

At the end of 1954, the tenth army-wide cryptographic conference was organized in Hanoi. The conference carried out a review and estimation of the situation involving every aspect of the cryptographic task during the time past and discussed the direction and responsibilities of the army cryptographic branch in the coming stage. The conference unified a number of problems concerning cryptographic technique and the professional task with important instructional significance:

1. Cryptographic technique is the product of the class struggle. It serves the Party and the army. With all their hearts and all their intellect, cryptographic cadre and personnel must serve unconditionally, determined to protect the essential secret matters of the Party in each difficult circumstance.

2. Cryptographic technique has three basic principles: constant secrecy is the first principle - speed and accuracy must be combined, and secrecy must be spread over the foundation.

3. Cryptographic technique serves the army, thus it is consistent with strategic and tactical thought, with the operational line and the form of the battlefield.[2]

The conference also made clear: "Secrecy, speed, accuracy are the basic content of cryptographic technique. We swear with all our hearts and all our intellect to stand definitely on the position of the worker class, always raising the technical level, ensuring the secret matters of the Party, the army, the people, in any circumstances at all, even though one must give his life to protect the secrets of the Party, we also completely safeguard honor and dignity. This is technical thought, or, to put it differently, our cryptographic professional thought."[3]

****

The People's Army of Viet Nam cryptographic branch was formed immediately after our people's democratic nation was born, amid conditions of difficulty and complexity. The initial cryptographic organization was established and received responsibility both for building organization and technique and for serving the great resistance against the aggressive French colonialists, to secure and keep the independence of our nation.

Through nine years of protracted resistance, following the teaching of Uncle Ho: "Cryptography must be secret, swift, accurate. Cryptographers must be security conscious and of one mind." The army cryptographic branch strove to go from have-not to have, at each step building and growing in every aspect, ensuring completely and outstandingly its responsibility for guarding the secrets of the Party and of the army.

From a few cadre receiving from the army the responsibility for the cryptographic task at the outset, until the success of the resistance, the army cryptographic branch built and developed a body of technical cadre and personnel to meet the requirements of serving the resistance. These were comrade cadre, party members, tempered and tested on the fields of battle and who had grown in their level of specialized technique, profoundly alive in revolutionary ideals, firm in class outlook, quality, virtue, well prepared to accomplish the responsibility that had been entrusted. These comrades were the precious capital in building and leading the branch upward to respond to the revolutionary requirements in the new stage.

From a position of having no capital at all from the standpoint of knowledge and cryptographic technique, the army cryptographic branch had searched out and created the science of Vietnamese cryptographic technique, with a technical level that never stopped advancing and rising, rendering unsuccessful the enemy's schemes for collecting information through cryptanalysis, protecting the secret content in the tasks of Party leadership, direction, and command of the army passed through the various communication media. While concentrating on the responsibility to serve the resistance against France, the army cryptographic branch laid the initial foundation on which to construct the basic theory of the science of Vietnamese cryptographic technique.

Through practice, the organization of cryptographic technique and the professional task methodology of the army cryptographic branch served to ensure command, conforming with the peculiarities of the army's tactical situation, raising step by step the level of military, technical, and administrative capabilities of the branch, overcoming initial dislikes and proceeding to build the cryptographic branch's professional specialty task activities in a regular manner, building each task relationship according to the hierarchical cryptographic system, fixing mission responsibilities clearly: tables of organization and regulations fixed, tight, consistent with the nature of the mission.

Although there were shortcomings and weaknesses in the technical professional tasks, principally in conditions at the beginning of building the branch (such as building the tables of organization, settling professional [matters] appropriate to the nature of the mission albeit still not timely; and the matter of expanding the level of technique, while still falling short of the requirements of the resistance) but through practice in the task of serving leadership, direction, and command, the accomplishments of the army cryptographic branch had been demonstrated and continued to be rather great.

The deciding factor in the spread, growth, and accomplishments of the army cryptographic branch was the leadership of the Party and the command echelons of the army. Concern on the part of the Central Party's Standing Committee, on the part of Uncle Ho, the coaching by the MND, the General Staff, the political commissars and commanders at the various levels--these were the sources of encouragement in strongly rousing the branch.

Bringing into play the accomplishments achieved, cadre and personnel in the army cryptographic branch strove directly to move up to accomplish successfully each mission in the new phase of the revolution: the phase of building socialism while resisting America, to save the nation.

Notes

1. The General Staff's order mobilizing military reeducation (1953).

2. Document of the Tenth Army-wide Cryptographic Conference.

3. Ibid.

Chapter Five

The Army Cryptographic Branch
Expands in Every Aspect; Widespread Use
of the KTB Technique; Participating in the Discharge
of Duties in the New Stage of the Revolution (1955–1965)

The war of resistance against the French colonialist aggressors ended in victory. Along with the entire military, the army cryptographic branch entered a new period of development, a period of building a revolutionary army, advancing gradually, step by step, into becoming regular and modern. In accordance with duty requirements, during this period the army cryptographic branch had to ensure service to guidance and command in the new conditions according to modern operational procedures, while at the same time building the branch, to progress toward becoming regular, with respect to tight organization, advanced technique, and a strict and clear regimen for the cryptographic task.

In August 1956, the General Staff organized the eleventh army-wide conference of cryptographic cadre, aimed at thoroughly grasping the situation of the new mission and coming out with a course of action for building the branch in the coming stage. The conference came up with three principal duties, namely:

1. Continue in-service professional reeducation and arrange for training in classes of appropriate levels, aiming at raising professional ideology for cadre and personnel with love and dedication toward their duty responsibilities, while at the same time raising the level of technique and principles of employing technique in order to ensure the essence of the cryptographic task.

2. Apply the KTB technique throughout the army and raise the technical level to a modern standard, placing special importance on solving the problem of mixed key at a sound level of technical requirement, and defining tightly the principles for the use of technique, while, at the same time, researching and preparing types of technique for use in operations in accordance with the army's line of preparation for combat.

3. Build a system for the cryptographic task in all categories, and plan to bring about branch-wide unification in the [cryptographic] field, organizing a system of monitoring and tightly controlling the implementation.

Reviewing accomplishments in the two years, 1954 to 1956, the cryptographic organizations army-wide had ensured secrecy, accuracy, and speed for the contents of leadership, direction, and command vis-a-vis the responsibilities of implementing the cease-fire order; concentrating the army and regrouping in the North; taking over the capital of Hanoi; taking over the 300-day zone; countering forced evacuation to the South;wiping out bandits in the Northeast and Northwest; eliminating the enemy's stay-behind reactionary gangs and the enemy commando-spies spread through the North; and implementing land reform throughout the North, etc.

In the 1956 army-wide conference of cryptographic cadre, Cde Hoang Van Thai, Deputy Chief of the General Staff, expressed his appreciation: "In fulfilling the heavy, difficult, and complex duties of our military and people, the army cryptographic branch has gone all out – its comrades are the people who have communicated secret instructions from Central and the Main Military Committee down to lower echelons and situation reports up from below, working so that these orders and instructions were fully comprehended, from South to North, over all the theaters. The extremely tense situation back then demanded timeliness by the hour – by the minute – and, first and foremost, to ensure absolute secrecy, and the comrades fulfilled that duty. Truly, the significance of that was extremely great."

From 1955 on, army cryptographic was realigned in accordance with the chain of command, from the MND-High Command down to the basic units. Many new cryptographic organizations in turn were established. Army cryptography expanded many times over, compared to the period of resistance against the French.[1] The General Staff Cryptographic Bureau proposed to appoint tens of cadre-in-charge and to supply hundreds of cryptographic cadre and personnel[2] in order to augment and strengthen the Military Region (quan khu-MR) and division cryptographic organizations, especially the newly established cryptographic organizations, such as the cryptographic sections of the 335th, 328th, 332nd, 330th, and 324th divisions;[3] the Western Mission group; the convalescence groups; the cryptographic mission teams with the Cease-fire Commission in the regions; the Cryptographic Section in the Directorate of Civil Aviation; MR 3, MR 4, MR Viet Bac, and Central Party Cryptographic; and supplementing the printing plant personnel. The adjusting and arranging of cadre-in-charge was reconciled with mission needs and level of ability of each person. The Cryptographic Bureau of the General Staff organized two additional sections, the Telegram Management Section and the Administrative Section.

As of the end of 1956, the system of organization of army cryptography comprised the Bureau of Cryptography of the General Staff as the lead professional organization of the branch, directly guiding the cryptographic task throughout the military. In the general directorates, MRs, divisions [su doan] and equivalent units, there was a Cryptographic Section. In regiments and small units there was a subsection or cryptographic team.[4] [This marks the 1955 change in terminology from the wartime term *dai doan* used in the preceding paragraph, to the conventional Vietnamese word for division, *su doan*, used hereafter. The change evidently was part of the "regularization and modernization" movement that also introduced rank and insignia and resulted in divisions of three infantry and one artillery regiments plus supporting battalions and companies. – Tr./Ed.]

Each year the Army Cryptographic School's classes training personnel enrolled more than the year before, students going about their studies enthusiastically and with a sense of urgency. In 1955, ninety comrades were trained; in 1957, a class of ninety-two comrades; in 1959, 307 comrades; and 143 comrades in 1960. School graduates were supplied as augmentees to bring unit organization and numbers up to strength army-wide. "An army proceeding to regularize and modernize is in much need of cryptography—needs many additional, skillful cryptographic personnel: even with machines there must still be people, people with a solid political stance, high class-ideals, skilled in technique, with a sense of dedication and endurance in performing the task of encrypting and decrypting."[5]

In order to improve the quality in the ranks of cadre and personnel, and to bring the organization up to strength, the branch opened many professional reeducation classes. In the year 1954–1955 there were 358 instances in which cadre and personnel took part in in-service professional reeducation and 245 comrades who took professional refreshers in school.

In 1956 the cryptographic branch organized professional refresher classes, one class for thirty-five comrades previously assigned in Intersector 5, one class for forty-five comrades previously assigned in Nam Bo, one for thirty-three cadre and personnel in various odd stations and nets, and one for KTB refresher and culture for seventy-five comrades.

Also that year, army cryptographic branch soldiers and cadre studied professional politics, task arrangement, and work style of cryptographic cadre and personnel.

Through study, criticism, and self-criticism, each cadre and person grasped thoroughly and profoundly their revolutionary responsibility and the responsibility of the army in the new phase, constantly displaying alertness and a willingness to fight on the front of keeping command secrecy through cryptography, opposing the enemy's schemes for collecting information through cryptanalysis, implementing regulations and speciality knowledge.

Understanding the three principles of cryptographic technique and general methods of applying the relationship of secrecy, swiftness, and accuracy, cadre and soldiers of the army cryptographic branch gradually overcame unsound perceptions and thoughts, calm and content in the task, prepared to carry out the responsibilities they had been given.

ESTABLISHING THE CRYPTOGRAPHIC SECTION OF CENTRAL AND THE CENTRAL PARTY SECRETARIAT, PROMULGATING TASK REGULATIONS FOR THE VIETNAMESE CRYPTOGRAPHIC BRANCH

On 21 July 1956 the Central Party Secretariat issued Decree No. 10-NQ/TW[6] establishing the Central Cryptographic Section, with these responsibilities:

- To help the Central Party research, oversee, and lead the cryptographic task in the regions and branches.

- To make plans for cryptographic assignments and oversee and guide the execution of the plans of assignment.

- To research and oversee the cadre and personnel situation, and to look after the matter of raising the political, cultural, and professional levels of the cadre and personnel.

- To be directly involved in assisting the cryptographic organizations of Party, government, and army at Central with respect to the profession and the problems deriving from the principles and regulations of the cryptographic task.

- To be directly involved in research and production of cryptographic systems for the branches and organizations of the Party, the government, and the army, guiding and assisting the cryptographic organizations in executing the work of selecting, developing, and refreshing the cadre and personnel.

- To research and propose to Central the promotion, appointment, and branch transfer of cadre and personnel.

- To inspect the implementation of the cryptographic regulations in the regions and branches.

- To report to Central concerning the task of Central cryptographic organizations and the situation of the cryptographic task in the regions and branches, etc.

The Central Cryptographic Section was placed under the direction of Central, under the direct charge of Cde Hoang Anh, committee member of the Central Party.

As for cadre in the Central Cryptographic Section, the Secretariat (Ban bi thu) decided as follows:

Cde Le Thanh Hai, Chief of the General Staff Cryptographic Bureau, was to be acting chief of the Central Cryptographic Section.

Cde Nguyen Manh Hoan, Chief of the Cryptographic Bureau of the Central Party Secretariat [Van Phong], was to be deputy chief of the Central Cryptographic Section.

After the August Revolution, the cryptographic branch of the army, the cryptographic branch of the Party and government, and the Public Security cryptographic branch were established in turn to serve the leadership and direction of the Party and government and the command of the army by means of cryptographic technique via communications means. In the course of building and working, the cryptographic branches had cooperated in helping one another in various aspects: cadre and personnel, cryptographic material, and professionalism, concerning technique and task experiences. The army cryptographic branch especially contributed positively in the building of the ranks of cadre and personnel and the routinizing of the technique task and professionalism of the fellow branches, until there came about unified directives concerning professional technique from the Central Cryptographic Section, opening up conditions for building the branch in the new stage.

On 17 December 1956, Cde Chief of the General Staff issued Decision No. 676/G8-TC detaching the Research Section and the Printing Plant of the General Staff Cryptographic Bureau to the Central Cryptographic Section.

Based on the nature of the cryptographic task and mission, in order to guarantee the unification and concentration of direction and ensure the secret tasks of Party and nation, on 20 November 1958 the Central Party Secretariat issued Circular #178-TT/TW, promulgating the "Task Regulations for the Cryptographic Branch" in order to bring about unification in the branches using cryptography nationwide.

These task regulations comprised nine chapters, ninety-six sections, which clearly defined the mission essentials of the cryptographic branch, general principles, principles of organization; regulations for cadre and personnel; regulations for technique research and allocating the use of techniques; regulations for encrypting and decrypting; regulations for electrical transmissions; and regulations for inspection and responsibilities of executive committees and chiefs of organizations (units) using cryptography or relating to the cryptographic task.

After getting the task regulations, the army cryptographic branch made a plan of study for cryptographic cadre and personnel army-wide, aimed at thorough comprehension of the objectives and meaning of the promulgation of the task regulations and to discuss ways in which to implement seriously and strictly the principles and definitions of these task regulations.

Based upon the Party's guidelines for strengthening organization in order to ensure the successful realization of every respect of the concrete guidance for the cryptographic task throughout the nation, and in accordance with a decision of the Main Military Committee and Central Cryptographic Section, the Central Party Secretariat [Ban bi thu] resolved to delegate to the Main Military Committee helping Central guide every aspect of the cryptographic task nationwide and to closely administer the cryptographic organizations of central.

Cde Hoang Van Thai, member of the Main Military Committee and deputy chief of the PAVN General Staff, was assigned by the Main Military Committee the responsibility of providing guidance to the Vietnamese cryptographic branch.

"The Cryptographic Section of Central and the Cryptographic Branch of the General Staff are united in producing an organization to guide the cryptographic task, with the mission of assisting Central and the Main Military Committee via the cryptographic bureau of the Central Secretariat and the Cryptographic Bureau of the Ministry of Public Security Secretariat in order to guide the cryptographic task of the Party and the government and to directly guide the cryptographic task of the army."[7]

Thus it was that the professional organizations of the Cryptographic Bureau of the General Staff had to carry out responsibilities vis-a-vis the army cryptographic branch while helping the Cryptographic Section of Central carry out responsibilities vis-a-vis the cryptographic branch of Viet Nam.

On 24 January 1959, the Naval Directorate established the Naval Directorate Cryptographic Section[8] under Cde LTjg [thuong uy] Vu Bao Phong, a comrade deputy section chief, and five cadre and personnel directly subordinate as Cryptographic of Base 1,

Base 2, Cryptographic of the school, and of a number of boat units and islands under construction.

The Navy's responsibility for action is on the battlefield of river and sea. The requirements of leadership and command are very broad and increase daily; the means of liaison in use is essentially by radio. The system of cryptographic organization expands down to unit level, in accordance with requirements for expansion and growth on the part of the naval forces.

On 10 March the Cryptographic Team of Group 130 was formed;[9] on 22 December 1959 the Cryptographic Team of Group 135 was formed. In April 1960, Naval Directorate Cryptography was augmented by fifty cadre and personnel. Afterward the Navy accepted from MR Left Bank and MR 4 the transfer of the system of the string of islands--Long Chau, Cat Ba, Co To, Vinh Thuc, Hon Ngu, Hon Mat, Hon La, and Mui Si. The number of naval cryptographic cadre and personnel rose to nearly 150 comrades.

In March 1959, Cryptographic of the Armed Public Security [Forces] was established. The Main Military Committee and the General Staff entrusted the Central Cryptographic Section and the Cryptographic Bureau of the General Staff with responsibility for building a system of organization and cryptographic-liaison network for Armed Public Security from HQ down to border defense sectors, isolated posts, maritime units, etc.

Initially, sixteen army cryptographic cadre of MR Viet Bac, Northwest, Left Bank, MR 4, divisions 350, 316, etc., were posted to Armed Public Security. Cde 1st Lt Hoang Quyen was decided upon as chief of the Armed Public Security Cryptographic Section. From the end of March 1959, the forces and cryptographic organizations of Armed Public Security took shape in the critical localities.[10]

In order to serve direction and command of the military transportation group supporting the South, in May 1959 the Cryptographic Section of Group 559 was established. Cde Nguyen Duc Mai, in charge of the Cryptographic Section of Group 559, received the mission of building the cryptographic organization of the Truong Son ["Ho Chi Minh Trail"] troops.

Afterward, in September 1959, the cryptographic system [he thong] of Group 959 was expanded, the section chief being Cde Nguyen Ba Dzung. This group was responsible for organizing cryptographic liaison to serve the internal situation of the Specialists Group [doan chuyen gia] and our Volunteer Army fighting on the soil of our friends, while organizing and maintaining cryptographic liaison between the Group and the General Staff and liaison down to the Vietnamese specialist teams in the friends' provinces and districts, and to organize classes to train cryptographic cadre and personnel for the friendly Lao nation, and to help the friends in this task.

In 1959 the army cryptographic branch appointed in turn 279 cadre and personnel to open many additional cryptographic liaison nets for these units: cryptonet of Group 800,[11] cryptonet for Group 301,[12] organizing [cryptography] additionally for Air Defense HQ at ten radio stations subordinate to the 260th Regiment, 10th Regiment and newly established 280th Regiment, four stations of MR 4,[13] four stations of MR Left Bank,[14] two Thanh Hoa border defense stations,[15] four stations of MR Northwest,[16] eight stations for the Intelligence Directorate [Cuc Tinh bao], one station for the Air Force Directorate,[17] and again organized a bandit elimination net in Ha Giang [Province] of MR Viet Bac.[18]

On 20 August 1960, the Cryptographic Sub-Section of the 202nd Tank Regiment was established, afterward expanded to become the Cryptographic Section of the Armored branch. Cde Khong Trieu was designated section chief.

While continually preparing to serve the forces, the army cryptographic branch was also training to change over to new technique and ensuring the training of a line-up of cadre and personnel for the party and Government cryptographic branch. The Army Cryptographic School successively organized many different classes of instruction. These were refresher courses in changing from the use of KTA to KTB, combining the raising of the cultural level, for forty-five comrades, training four new classes, a class of thirty-five comrades for Party-Government Cryptographic, a class of sixty-nine especially for Armed Public Security, and two classes for the army.[19]

Compared to 1958, in 1959, army-wide, there were an additional sixty-six units using cryptography.

****

In 1955 the Cryptographic Bureau had a plan to provide guidance and expand the replacement of the cryptographic system army-wide, replacing [with] system KTB in the divisions [dai doan] and independent regiments, researching a system for the international net, researching a system for the Intersector and division nets, for units directly subordinate to HQ, the General Directorate of Supply, military intelligence [quan bao],and various odd nets, such as the Graves Registration Committee, Consultation X, and the HQ delegation group in Saigon.

The direction of the mission for the technical task of the branch was to change the direction for use of the KTB dictionary code from one-part [lit., half-mixed] to two-part [lit., completely mixed], especially in the divisions and main force units, implementing the use of KTB broadly at sector level. KTA was to be improved to serve basic units and directly subordinate branches.

Code books [lit., "cryptographic dictionaries"] are constructed as either one-part or as two-part, raising the degree of security and creating conditions for ease of production and use. The contents of the code books gradually became perfected, suitable to the vocabulary of leadership, guidance, and command of the High Command and of every organization and unit.

The Hoa Binh [PEACE] system had more advanced construction--from the end of 1955 until the beginning of 1956, it was put into use between HQ and the MRs, for main force units and the Cease Fire [Commission]. The Dzan Chu [DEMOCRACY] System and the Doc Lap [INDEPENDENCE] System were constructed according to the half-order principle [nguyen tac nua thu tu].

Random [loan] key also began to be researched and produced to achieve a high level of security.

Cryptographic security consciousness in production, allocation, maintenance, and use was also increased. Concerning technique KTA, the number of "compound words" in the code charts was expanded to enrich the content of the system. The structure of the key strips was unified and strict usage determined, overcoming part of the weaknesses of the KTA technique.

Regulations for use, mainly [cryptographic] key regulations, were improved, cancelling some complicated regulations that were not indispensable but were still ensuring secrecy. It can be said that this was a time in which the level of technique KTA to ensure secrecy, accuracy, and speed reached its peak.

With the above advancements, the types of cryptographic techniques met the requirements of handiness in use and protection of secrecy, faced with the increase in activities by the enemy to collect information through cryptanalysis. The army cryptographic branch elevated its technique exactly according to instructions from the Main Military Committee and the General Staff.

Moving into 1956, the system of use of technique KTB started to expand, from MR nets to regiments directly subordinate to the MR's, main force units, airfields, etc. – 14,352 sets of types KTA and KTB were produced in one year.

To implement the line of changing the technique, in 1957 many refresher classes in the new KTB technique were organized at HQ and in the MRs: two refresher classes in changing from technique KTA to KTB for 180 comrades; short, day refresher classes on changing technique in MR 4, MR Left Bank, and MR Viet Bac were undertaken by MR cryptographic sections while keeping quality relatively high and strictly implementing the content of HQ's plan.

Also in 1957 there were 170 comrades (mostly new graduates from school) sent by the Cryptographic Bureau to augment the units and MRs in expanding the changeover in technique. By the end of the year, more than 100 comrades had completed study of KTB and gone back to replace comrades using KTA so the latter could go to school for KTB training.

By the end of 1957 army-wide liaison nets (including internal MR nets) were using KTB, a major advance in the technique of the army cryptographic branch.

Also after 1957 the army cryptographic branch did not work on research into technique KTA, but concentrated on research and development and raising the level and productivity of technique KTB. All three components of this form of technique – codebook, random key, and principles of use – were considered. The content of the codebooks was made "richer," and as a result, more closely aligned with the command vocabulary particular to each unit. Random key was researched and produced according to new formulas and methods to attain a reliable level of security. Principles for use were closely defined. The "superenciphered, random numbers chart system," the "superenciphered, random numbers handbook," the walkie-talkie system, etc., were researched and began to be put into use, with results that built-up feelings of elation and enthusiasm on the part of cadre and personnel.

In 1957 the branch's printing plant received much additional equipment: a printing press, paper cutting machine, serrating machine, automatic type setter, etc., so the matter of printing was more favorable.

In 1958 the plant produced sixty-three types of the KTA and KTB systems, with tens of thousands of sets supplied to the army cryptographic, the Party, Government, and Public Security, comprising eight type-KTB two-part, eight telephone codes, ten codes for secret letters [mat thu], and eleven of the one-part type. KTA comprised sixteen of the spell-chart type, twenty-four specially made for cadre in independent action, two for Central for point-to-point liaison with two places. Production and distribution for the units was 1,777 sets of key of various types (army, 1,410 sets; cryptographic of the Central Secretariat, 269 sets; and Ministry of Public Security Cryptographic, 98 sets).

In 1959, the Techniques Research Section had completed research on ten new dictionary-type codes for the liaison nets of the army, intelligence [tinh bao], Party and Government, including a type of system used for liaison between Sector 5 and Nam Bo, researched eleven thin dictionary-type systems for the special liaison nets, with two systems for Group 301, three types for the Armed Public Security, Navy, and Air Defense border posts and observation stations, and one type of system for Group 959.

In addition, the army cryptographic branch made three more types of operations code [mat ngu] of the A-code form for the air defense observation stations and command cadre use (not embraced in the cryptographic system [he thong]).

In the use of the technique: Ensuring 100 percent accuracy in encryption and decryption was laid out as an essential requirement in overcoming adverse influences on guidance and command. The decisive attitude in the branch was that there could be no cessation in elevating the level of productivity above the basic 100 percent accuracy insurance, and that, above all, was the matter of ensuring secrecy. The entire branch was determined to strive to achieve the norm in real-life practice, not to have to reencrypt messages, not to garble the contents of secret messages so as to impact adversely on command. In training and practice, we had to carry out 100 percent accuracy in encryption and decryption. And we wanted to achieve that besides raising the sense of responsibility while still having to carry out precisely the practices and rules of encrypting and decrypting and checking thoughtfully after encrypting and decrypting.

The training task is always the most important task. The annual, quarterly, and monthly training plans became more adequate and more suitable every day. Training and study gradually became a mass movement. Voluntarily and enthusiastically, cadre and personnel participated in study of politics, military [matters], culture, technique, and professional [duties]. The results enthusiastically aroused the cryptographic cadre and personnel of the entire army to strive harder, to advance more each year than the year before. Many new records were set; the number of people achieving the norm gradually increased.

In 1955 the average productivity branch-wide in encrypting and decrypting with technique KTB was 350 groups per hour with an accuracy of 99 percent, with the 325th Division Cryptographic surpassing the entire branch with an average of 500 groups/hour.

Many individuals attained the record in productivity and accuracy. Cde Dinh Van An, a 325th Division cryptographer, encrypted 520 gps/hr and decrypted 758 gps/hr, with an accuracy of 99.50 percent. Cde Vu Hai, General Staff Cryptographic Bureau, encrypted 500 gps/hr and decrypted 700 gps/hr regularly, ensuring accuracy at 100 percent.

Average productivity in the branch, vis-a-vis technique KTA, was rather high: 320 gps/hr, 99 percent accuracy.

As a result of doing a good job with the ideological task, army cryptographic cadre and personnel clearly received the place and role of responsibility of party members performing the cryptographic task, and the ideology of never being content if the cryptographic task slipped a notch.

By the end of 1959 average productivity in encrypting and decrypting by the cryppies was 356 gps/hr vis-a-vis the nets using technique KTB.

The grand total of messages encrypted and decrypted by the units army-wide during the year was 271,436 official messages.[20]

****

Simultaneously with the strengthening of organization and the raising of technique, in two years (1954-1956) the army cryptographic branch carried out reorganization and getting the professional task on the right track. The internal regulations for the task promulgated and implemented from the tenth army-wide conference of cryptographic cadre in October 1954 were built upon to produce provisional task regulations and initially brought into play effectively. Cadre and personnel clearly sensed that "work style is a manifestation of one's ideology and service role – good work style is a concrete manifestation of the implementation of the task regulations."[21]

By mid-year 1955, the research, production, and allocation of cryptographic technique was unified and consolidated in the Cryptographic Bureau of the General Staff. Implementing decisions concerning the collecting of statistics, oversight, maintenance, use, retrieving and destroying by burning the various types of cryptographic systems, and the implementation of regulations and procedures for encrypting and decrypting messages became tighter, stricter, and more serious.

Cases of violating principles, such as using the same key for encrypting and decrypting, diminished. "Checking" after encrypting and decrypting was implemented, thus promptly detecting and correcting mistakes in work style and technique. The implementation of message regulations also clearly evolved in the cryptographic organizations as in the command organizations. The sense of vigilance on the part of cryptographic cadre and personnel was raised in activities and social relations.

Cryptographic organizations at the various levels were aware that "Because of the nature of the technical and professional task of the branch with respect to the content of the task, principally concerning the technical aspect, there is a close relationship from top to bottom and a ripple effect throughout the whole branch. One mistake, large or small, on the part of one unit or one region, with respect to technique, work style, or regulations can have adverse repercussions for the whole branch. One initiative, one experience of this unit can be applied to advantage by another unit and the whole branch";[22] therefore, there was an increase in the task of professional guidance by means of concrete measures. Cryptographic organizations above kept in close touch with cryptographic organizations below to aid them in building and the task of advancement; cryptographic organizations below did a good job of routine reporting to seek opinions and help from cryptographic organizations above, enabling them to grasp the situation thoroughly and more accurately, thus making guidance more to the point, and feelings of unity and affection in the branch became more pronounced, increasing additionally the power to fulfill the shared mission of the branch.

Each relationship between the cryptographic organizations at the various levels with unit commanders, operations organizations, and communications also became closer each day, so the organization to ensure command secrecy by cryptographic technique over the various means of communication made much progress.

For the years of building and working under peacetime conditions in a completely liberated North, the army cryptographic branch "fully completed its responsibilities in face of the historic situation . . . it absolutely contributed certain accomplishments and made clear progress."[23]

Aside from the accomplishments and gains, the army cryptographic branch still had shortcomings and limits in the results of the task and in the building of the branch. In the task of cadre organization, the branch had not yet fully grasped and correctly applied the Party's viewpoint concerning cadre policy, and with realignment and strengthening of organization, therefore, we had many cadre and personnel who, through training, were tested and stored up many experiences in the professional and technical tasks, so that, depending upon cryptographic task conditions, they could be transferred to other tasks or demobilized and discharged. As for the old cadre, the comprehensive upgrading had not yet been truly a matter of concern, so there were adverse influences on development with respect to branch organization and technique.

In January 1959, the fifteenth congress of the Party Executive Committee decreed: Except for the revolutionary path, the people of the South have no other road to escape the yoke of slavery, "the basis for development of the revolution in the South is the use of violence."

Central Party also decided "to liberate the South, to escape from the domination of imperialism and feudalism and bring about the independence of the race and give fields to the plowmen, completing the people's democratic racial revolution in the South."

The resolution pointed out the direction for building for combat on the part of the people's armed forces, creating conditions for our troops to have additional time and initiative to prepare in every respect, making ready to respond to the developing requirements of the revolution.

From the troop movement and regroupment until 1956, of the number of Cryptographic cadre and personnel remaining in the South most had been transferred to other assignments. A few continued the use of cryptography for the secret "white letter" [bach thu – open mail?] liaison line in order to ensure contact between the Nam Bo Regional Committee (Xu uy Nam Bo) with the various places and with Central.

After having the resolutions of the fifteenth [congress], implementing instructions from above, the cryptographic branch implemented the building of cryptographic organization in the armed forces in the South, so as to ensure command from Central and a cryptographic liaison net over the entire South. The Central Cryptographic Section and the General Staff Cryptographic Bureau assembled cadre and personnel who were native to the South and regrouped in the North, and built them up in every respect, preparing them to be ready to return to the South on assignment. The comrades who received this important mission and honor were very enthusiastic and proud of the trust of the Party and army, eager to study and thoroughly comprehend the way, the revolutionary mission, to grasp the technique. Early in September 1959, Cdes Tung and Pho went with a cadre group down to Sector 5. Starting out from Relay Station 354 by automobile, they arrived on the north side of the Ben Hai river, then continued on foot. After nearly two months of climbing mountain passes and fording streams, on the march during the day, resting at night and encrypting and decrypting contacts with HQ, the comrades arrived at the Sector 5 Sector Committee at the end of October. During this time many cryptographic cadre and personnel went down south along with groups of Party cadre, leadership cadre, and army cadre, going down to perform their missions in the South.

Even though distant from Central's professional guidance, the cryptographic cadre and personnel in the South overcame so many difficulties and hardships in order to serve, bringing up a spirit of industry, self-reliance, economizing with every scrap of paper, every drop of ink, etc.

In the years of peacetime activity, under the guidance of the Main Military Committee, the Ministry of National Defense, and the General Staff, the army cryptographic branch had new progressive developments.

From 1954 to 1960, the number of points in contact increased sixfold, the volume of messages encrypted and decrypted increased fivefold; the ranks of cadre and personnel never ceased to grow stronger in terms of quantity and quality. Almost all cadre and soldiers were party members of the Vietnamese Lao Dong Party. The army cryptographic cadre and personnel comprised nearly 70 percent of the total cryptographic cadre and personnel nationwide and more than 60 percent of the units nationwide having cryptographic liaison.

The task of research to develop the science of cryptographic technique and to produce professional technical means of ensuring the secrecy of command loomed large, in order to counter the enemy's cryptanalytic tricks and schemes. The need to produce the types of systems, cryptographic key and other materials, along with the distribution of guidance on usage throughout the military increased from day one. The system of cryptographic organization in the MRs, services [quan chung], and branches [binh chung], and that of Armed Public Security cryptography was brought up to strength and expanded very rapidly.

Faced with these urgent needs, in March 1961 the Ministry of National Defense decided to transform the General Staff Cryptographic Bureau into the General Staff Directorate of Cryptography. Organizations of the Cryptographic Directorate were brought up to strength with respect to responsibilities, mission, and tables of organization. Cde Le Thanh Hai was appointed chief of the Directorate, with Cde Nguyen Dzuy Phe as deputy. Organization of the Directorate and units directly subordinate comprised

- The Bureau of Cadre Organization, under Cde Nguyen Chanh Can,

- The Bureau of Technique Research, under Cde Le Van Bang,

- The Bureau of Message Encrypting and Decrypting, under Cde Luong Van Tin,

- The Army Cryptographic School, with Cde Pham Tu Cap as commandant,

- The Printing Plant, with Cde Chu Van Hoan as director.

Thoroughly grasping the mission of building the ranks of army cadre and the realities of the cryptographic branch, the Cryptographic Directorate worked with the Directorate of Military Personnel to prepare cryptographic tables of organization for the MRs, services and branches.

The Directorate decided that, in order to resolve long-standing deficiencies in manning, units having missions in Theater of War C [Laos], such as MR Northwest, the 351st Division, the 367th Division, etc., would receive supplemental personnel.

Of the students who were graduated from the training class at the Army Cryptographic School, ninety-three comrade personnel were allocated to units in the North: Air Defense, Air Force, Navy, Engineers, Artillery, MR Left Bank, MR Right Bank, MR Viet Bac,MR 4, and units directly subordinate to HQ. The organization of a reserve element of the Directorate, comprising thirty-six people (thirty-three of them noncommissioned officers), among them newly graduated students from the school, was prepared to augment the units, doing work that had just popped up or replacing cadre and personnel in units with people off studying in professional refresher courses. Because of the reserve forces, the Directorate was able to augment on a timely basis and serve the campaigns well.

The printing plant also selected an additional number of comrades, among them a number of soldiers who had completed their military service in the units, in order to ensure production of [cryptographic] materials. In the first six months of 1962, the printing plant of the Cryptographic Directorate exceeded the planned production by 9.5 percent for cryptographic systems and professional papers.

The task of serving the assistance planned for the South was carried out zealously. After reexamining the situation involving the groups on operations and the means of professional [cryptographic] materials that had been sent to the South, the Directorate issued a communique to Central Office cryptographic concerning the situation of augmenting cadre and personnel up to that time, after which they implemented a summarization of the cryptographic task of ensuring service to Theater of War B [South Viet Nam]. From the middle of 1961 to April 1962, Northern Cryptographic assisted the Southern theater of war with 287 comrades, among them 118 people for Nam Bo cryptographic, 90 for Sector 5 cryptographic, 41 for Sector 6, 56 for Intelligence [tinh bao],12 for [Quang] Tri- [Thua] Thien, and 17 for Group 559.

After an army-wide cryptographic professional conference (in 1961), the Directorate corrected the documents and guidance for concrete implementation of the contents of conference resolutions and conveyed this down to the units. Vis-a-vis the large units, the Directorate sent cadre down to publicize this in person.

So as not to flag in efforts to raise the level of cryptographic cadre and personnel, the Directorate issued guidance and made plans for units such as the Navy Cryptographic Section, the cryptographic sections of the Left Bank and Right Bank MRs, the Cryptographic Section of the 367th Air Defense Group, and the Cryptographic Section of Armed Public Security to start professional training classes.

The Army Cryptographic School essentially trained personnel for the military regions and divisions. The classes achieved lofty results in practice, with respect to productivity and degree of accuracy, the highest productivity in encrypting being 225 gps/hr by book code and 331 gps/hr by chart code. For decrypting, the highest was 321 gps/hr by book code and 382 gps/hr by chart code.

The training classes for Theater of War B also achieved good results. One class of forty people was especially for Military Intelligence, two classes of eighty-seven and sixty-one people for Sector 5 and the Highlands, and a cryptographic organization from Sector 6 to go back down south.

Harking back to the South of their birth, cryptographic warriors trained, studied their specialty and studied politics day and night, patiently enduring military studies, enduring realistic training, training in carrying heavy loads over a long distance, preparing to go down the Ho Chi Minh trail into the South to strike America.

The Techniques Research Bureau in the first six months of 1962 completed research on five types of systems for the Central Office, nineteen for the services and branches in the North, ten for preparing assistance for the units in the South and six types for units in different theaters. An element of Research Bureau cadre finished 30,000 units of key of various types and analyzed and rechecked it for accuracy prior to putting it into use. Besides all of this, cryptographic key models of the five- and ten-column type were modified.

As for the task of instruction in use, the functional organizations of the Directorate issued cryptographic key and replaced cryptographic systems for many units, with 2,956 sets of cryptographic key used for the lateral contact form, skip-echelon contact, and combined operations contact, issuing thirty-two types of systems, consisting of 759 sets, promptly ensuring the requirements of the units (the campaign liaison net).

At the beginning of 1962, the Directorate organized a summarization of the situation of rectifying technique and ensuring cryptographic secrecy throughout the nation. Through this would come a standard for equipping cryptography at the various levels.

Also during the year the Directorate organized an inspection group at the cryptographic work places of a number of units (MR 4, comprising MR HQ Cryptographic, Cryptographic of the 325th and 341st Divisions, and Armed Public Security Cryptographic in Vinh Linh, Quang Binh, and Nghe An). Directorate cadre went down to basic units to inspect the management of technique and grasp the materials of the liaison nets of the 305th Division and of MR Northwest (comprising the 316th and 335th Divisions). Through inspection, the Directorate could clearly see the situation involving the cryptographic task, with respect to staffing, organization, ratio of cadre to personnel, and the concrete regulations covering cryptographers on the job (both from the standpoint of material and spirit). Also through inspection cryptographic cadre achieved a meeting of the mind with unit commanders and cryptographic sections concerning the relational task involving command guidance and service organizations, producing improvements.

In the Message Encrypting and Decrypting Bureau, the liaison net was arranged on a wide sphere with many elements, consisting of units having responsibilities within the nation and others having international responsibilities, among these units relatively fixed and units continuously on the move.

In 1962 the number of places in liaison with the bureau was 112; in 1965 it was up to 370. With the increase in number, the bureau still grasped thoroughly the principles of arranging contact and the professional equipment to ensure the work of encrypting and decrypting with the units. Besides the system [he thong] of cryptography used for regular communication, there was a system of cryptography that anticipated a situation of sudden expansion in order to serve joint liaison, be in reserve, or be used for special nets. Never letting up in doing a good job of performing the principal mission of encrypting and decrypting, the bureau still did a good job of performing the tasks of receiving and transferring, arranging, collecting statistics, logging, extracting and closely following the situation involving usage, and the concrete requirements of each unit in the theater, giving timely help to the Directorate in guiding the task of cryptographic technique usage.

As the center for connectivity – for accepting and forwarding contact with the units, services, and branches throughout the army, many points consolidated and the liaison net expanded – it was usually not fixed; in 1965 alone the bureau used fifty-five types of systems to encrypt and decrypt 105,846 secret messages of 8,933,449 groups without any major error adversely influencing upper echelon guidance.

Thoroughly grasping the Party's 1961–1965 resolution on the military mission and the second five-year military plan, resolution #60/NQ-TW, 17 November 1962, from the Central Secretariat realigned the organization of the Central Cryptographic Section and the cryptographic organizations of the Party, the administration, and the army branches, the army cryptographic branch having to step up its task of strengthening and expanding the system of organization in the North, while at the same time building the system of organization in the South so as to satisfy every requirement of leadership and command by means of cryptography.

The system of organization of the army cryptographic branch expanded all over the two areas, South and North, as well as the Laotian theater. In the North, the army cryptographic system expanded an additional step. From 1963, the Cryptographic Sections in the military regions, arms and General Directorate organizations, and equivalent units were brought up to strength and turned into cryptographic Bureaus, directly subordinate to the staffs [Bo Tham Muu].

After the establishment of the Air Defense-Air Force service [quan chung] in October 1963 the service's Cryptographic Bureau was established, comprising the cryptographic organizations of the air defense, air force, and radar branches [binh chung]. Cde Pham Dzuong was appointed bureau chief. Based on the organization of forces of the service and the basic operational plan, the Cryptographic Bureau organized a cryptographic liaison net in the entire service and joint liaison with friendly units.

The Air Force cryptographic net was arranged according to the command system in the lines of airfields, comprising the system of permanent airfields (first line airfields from that class up), the system of reserve airfields, and the system of field [da chien] airfields. Following conditions and realities of combat service, the cryptographic nets were officially arranged in the permanent airfield system and the reserve airfield system, while for the field airfield system, its cryptographic net was arranged according to the mobile radio station model, according to the mission requirements of each battle for specific arrangements.

[Air] regiments had lateral contact with each other in regional coordinated combat. All permanent and reserve airfields had arrangements for skip-echelon contact with HQ.

The forces of the antiaircraft, rocket, and radar branches expanded rapidly. When massed for a strike, there were AA and rockets, as well as the air force, in coordinated combat, but when protecting [lines of] communication, they were dispersed by battalion on lengthy highways according to main points for which the regiment or division was charged, and stretched out over hundreds of kilometers. Around the end of 1963, there were more than ten AA and radar regiments, but still carrying the name of the 367th Group, the Cryptographic Section of Group 367 comprising the chief and ten cadre and personnel. The AA regiments had a table of organization with a subsection of five to ten cadre and personnel. The radar regiment had eight to nine cadre and personnel; each radar company had one person for encrypting and decrypting. Cryptographic at the regimental level had a commissioned officer in charge. Normally this officer comrade had combat and command experience and a high sense of responsibility. Thus they built in an orderly manner the management and tight organization of their elements, implementing the mission well, building up confidence on the part of the people in command,

From 1963 to 1965 the technical branches of the service expanded more and consolidated their strength more. Rocket and AA regiments and air defense divisions were established to guard the various sectors. The cryptographic sections of these divisions took shape and became fully worked out day by day.

Encrypting and decrypting is the work of ensuring service day after day to the direction and command of the service, having closely adhered to the requirements of the service in every battle.

During this time the units of the service had numerous means of rapid communication-liaison, not much message volume going via cryptography, but messages regularly having high precedence (the latter making up 60-70 percent of the total). Air Force messages were regularly sent out from 1800-2000 hrs. the day before until 0500-0600 hrs. the day after execution. Preflight messages, meteorological forecasts, transport service, assistance to our friends. Work was concentrated, for the most part, at nighttime. The fellows in cryptography understood quite clearly that, to serve a modern service and branch, the task of leadership and command via cryptography carries importance and much urgency. The duration of operations is figured in seconds – in minutes – but the task of preparation demands the very best. That applies to the exact time period for much of the cryptographic task.

In the campaign to protect the city of Vinh in 1965, the staff of the service [Bo tham muu quan chung] requested the transmission of orders by encrypted messages with the time from issuance of the order to unit receipt being five minutes. The service cryptographic bureau organized a shift of comrades with a rather good specialist attitude to go to the CP to work directly in reading or listening to the order and encrypting it at once, at the same time decrypting and reporting directly to the command cadre. The results achieved the command requirement for timeliness set by HQ, victoriously striking the enemy in the campaign, and received commendation by the staff.

****

In order to be compatible with the conditions and essence of the task activity of the Naval service, the service cryptographic organization organized short day-training classes. The study content comprised study to thoroughly grasp the mission situation of the service and the branch, generally speaking, and, specifically, the responsibility to build navy cryptography. Navy cryptographic also made efforts to organize and straighten out their ranks, to make the cadre and personnel love the branch of work, to become attached to the service, to the sea, to the islands.

In 1962, cryptographic of the 2nd Patrol Sector, Naval Base 2,[24] serving the unit striking the American-Puppet commando boats encroaching upon the Central region coast, used technique KTB and a command operations code [mat ngu chi huy].[25]

On 11 August 1962, Petty Officer Third Class Bui Dang Dzuong, a cryptographer at Naval Base 1, received an order to go out to Long Chau island on assignment. When the boat went past the zero buoy, it met a misfortune. In big waves and heavy wind, the comrade continued firmly at the con, unruffled he rowed and steered, and, together with his mates, bailed water from the boat. The misfortune continued. Wind and wave on the sea [suddenly] large, the boat being small, the mast and the oar locks snapped, and the boat capsized. Before the boat sank, Cde Dzuong calmly destroyed the entire set of [cryptographic] materials and continued to encourage and help his mates get out of the boat and swim to the island. Big waves, heavy wind, and sapped of strength – Cde Dzuong gave his life.

In the Three Firsts emulation movement, May 1962, Navy cryptographic organized the first technical competition and exhibition meet at Do Son. Through the meet, many forms of study or skill drills of the service's cryptographic [organization] were publicized in the branch.

After the 3 January 1964 decision to establish Naval HQ [Bo tu lenh Hai quan], the Cryptographic Section of the Naval [Directorate] became the Navy Cryptographic Bureau, under Cde Vu Bao Phong. The encrypting-decrypting, message, and technique elements of the bureau had veteran cadre in charge, with the capacity to respond to the service's leadership and command requirements for combat at sea.

In January 1964, the Cryptographic Section of Group 125[26] was formed to serve the unit transporting weapons and munitions, means, and assistance forces for the Southern theater. Cde Nguyen Duc Bao was section chief. The ranks of cryptographic cadre and personnel chosen to go down south by boat were comrades with high sense of responsibility and valor.

MR 4, as the front line unit of the North, directly in touch with the enemy at the temporary demarcation line, had a long border and coastline. MR cryptographic organizations had to ensure combat readiness for internal units, while at the same time having to cover our action and that of our friends in Theater of War C. That situation meant that MR cryptographic tables of organization could not be fixed.

In order to consolidate plans for defense and for combat readiness and to arrange cryptographic materials, MR cryptographic proposed to the Directorate the organization of a meeting comprising MR cryptographic and divisions and brigades subordinate to the MR, with Armed Public Security units, MR Right Bank cryptographic, and naval and antiaircraft [units] stationed in the MR's sector of responsibility.

In initial combat service, MR 4 cryptographic encountered a number of difficulties, but, through the direct professional guidance of the Cryptographic Directorate, MR cryptographic overcame a number of shortcomings and accomplished the mission. The armed forces in the MR expanded rapidly, and the use of cryptography also increased. In 1964, the MR had thirty-four units using cryptographic, and in 1965 that had gone up to seventy-one units. At times they had to assign four to five assault [radio] stations for engineer and AA units, etc. The cryptographic cadre going on independent missions were excellent, and had recently ensured timely encrypting and decrypting of messages and corrected errors in the handling process. Therefore the MR cryptographic bureau had to divide up the number of people remaining at the bureau to create two to three elements to work. Three comrades worked the message task day and night, replacing each other in receiving messages, sending them off, making copies, getting them back.

Confronted by the lack of cryptographers, the MR HQ, the comrade chief of staff, the military personnel and guard organizations, and the Party committees at the various levels assigned cadre to go down to units to select people to come to HQ to study. But by the end of 1965, the MR had only accomplished 70–75 percent of the troop strength HQ provided. The MR cryptographic bureau and a number of division cryptographic sections, such as those from the 341st and 325th divisions, took advantage of the time for professional replenishment for their cadre and personnel, so that they would have sufficient capacity to perform the task of encrypting and decrypting independently and would grasp and accurately apply the principles for the use of technique and procedures to encrypt and decrypt a message. By means of the replenishment form of on-the-job training, in less than a year there were twenty comrades recently out of school who had taken on task responsibility in an isolated station or replaced the former comrades so they could go to different units. Summarized through the reeducation sessions resolved by Central and the MR Committee, cadre and personnel raised their awareness of the new mission situation, the mission of liberating the South, of guarding the North, of supporting the Lao revolution. Many comrades rushed to go receive missions in places most difficult and arduous, or far distant places, such as Hon Co, in AA units, or in Theater of War B or Theater of War C.

In Theater C, Cde Cong Thanh was bombed by enemy planes while working – his ears were deafened and ran blood; Cde Le Hong Qui, out on an operation, tripped an enemy mine that snapped his leg – two comrades calmly preserved the cryptographic systems and means for the task, and only when there was a replacement was he content to go to the hospital for treatment. Many comrades had fits of fever, yet when messages came they pushed themselves to encrypt or decrypt and would not let the messages be delayed.

On Hon Co, many nights there were hourly report messages, and cryptographic personnel had to stay awake continuously on many nights in order to promptly encrypt and decrypt so as not to lose any of the time element on which the message depended. In the 929th and 925th regiments, up in the mountainous forests of the Vietnamese-Laotian border, oil lamps, pencils and paper were inadequate for the job, and they gave their own money to go out and purchase.

In 1965 the MR 4 message volume went up by 191,684 compared with the official messages of the previous year. At MR HQ, in particular, there had been a peacetime daily of some eighty to ninety secret messages – in 1965 the average was 200 official messages a day, with days in which there was a three- or fourfold increase, and 70–80 percent were high precedence (Immediate and Priority) [TK va TGK].

During this time, service to leadership guidance and combat command in the two parts of our nation and the friendly Laotian nation was rather urgent, especially when the military and people of the South opened the counteroffensive against the American imperialists' "special warfare." Daily messages volume increased, carrying significant content concerning the line, resolutions, strategy, tactics, campaigns, stratagems and combat, etc. Cadre and soldiers of the entire army cryptographic branch heightened political responsibility and unity, determined to strive to complete the mission in every condition. Cryptographic organizations at the various levels attached special importance to the task of encrypting and decrypting messages, concentrating to exert themselves to the utmost in this important mission. One great difficulty and well known during this period was the lack of many cadre and personnel. The peacetime tables of organization were not compatible with the developing situation of revolutionary warfare taking place on a scale that grew larger each day. The urgent work of training and development, although having obvious effect, had many shortcomings in terms of basic requirements, thus was impossible to sustain over a long time.

From 1960, the revolution in the South underwent new advancements. In the spirit of "whatever's necessary, because the South is our brother," mission reinforcements for the Southern Region were increased.

In early May 1961, the Central Military Committee and the MND organized Phuong Dong ["orient"] Group 1 to go down to B2 and Phuong Dong Group 2 to go down to MR 5. In both of these two groups were elements of cryptographic cadre and personnel selected as augmentees to be the nucleus in creating cryptographic organizations for the various theaters. Besides the responsibility of the operation itself, the cryptographic cadre and personnel were also responsible for ensuring operational command liaison between the groups and the General Staff.

Comrade Tran Van Quang, Deputy Chief of the General Staff, took direct control over Phuong Dong Group 1, handing this responsibility to the Cryptographic Directorate and the cryptographic element in the group: ensure close cryptographic liaison between the group and the Central Military Committee and the General Staff during the troop movement; most especially, organize a cryptographic liaison system linking the [Southern] Region Military Affairs Committee with the Central Military [Party] Committee, with the General Staff, and MRs 7, 8, 9, and Saigon-Gia Dinh; upon arrival, set up a cryptographic system [he tong] for the Region's armed forces and the MRs.

The formation of a cryptographic organizational system [he tong] in the South was basically prepared from Phuong Dong Group 1 and Phuong Dong Group 2. The Cryptographic Directorate had thoroughly grasped the mission and zealously organized the implementation, concentrating on the training and augmentation of cadre and personnel and an increase in assistance with respect to technique, equipment, and professional means.

Tons of equipment means, cryptographic systems, and cipher key of various types were apportioned to each team, to be backpacked by each individual.

In Phuong Dong Group 1, there were twenty-nine cryptographic cadre and personnel. By the end of July, the group arrived at the regroupment position (the receiving station in Ma Da, subordinate to old War Zone D).

People and means were both safe. Throughout the itinerary the cryptographic cadre and personnel ensured timely and reliable traffic between the group and the Central Military [Party] Committee and the General Staff.

The cadre and personnel concerned with the cryptonet, together with the necessary means, were arranged to go at once to the military regions. In MR 7 (T1--the region of eastern Nam Bo) there were four comrades under 1st Lt Nguyen Ngoc Sinh. MR 8 (T2 – the region of central Nam Bo) had three comrades under 2nd Lt Nguyen Tuan Suong. MR

9 (T3 – the region of western Nam Bo) had four comrades under 2nd Lt Nguyen Van Duoc. The Saigon-Gia Dinh MR (T4) had three comrades under 2nd Lt Vo Xuan Tra.

The Cryptographic Bureau of the Region Military Affairs Section (R) and the units directly subordinate to Region had fifteen comrades under Comrade Captain Nguyen Hoang. After a few days' rest, the personnel moved the Region organizations down to Sector B (Trang Chien-Tay Ninh). The Cryptographic organization of the Region Military Affairs Section, carrying the designator [phien hieu] B8[27] had the mission of cryptographic liaison between the Region Military Affairs Section and the Central Military Party Committee and the General Staff. Each day an average of forty messages was encrypted and decrypted.

From the end of 1961 to the beginning of 1962, units directly subordinate to Region were established. The cryptographic comrades from the Region Military Affairs Section took turns going to the cryptographic units: Region Forward, under comrade 2nd Lt Dzuong Minh Tri; cryptographic of the Ben Tre Maritime Transportation Supply Station [Tram]; under comrade 2nd Lt Tran Minh Dat; the Ma Da area [vung] Rear Services Base, under Comrade Warrant Officer Pham Ngoc Linh; cryptographic of the U50 Guard Battalion under Sergeant Major Huynh Huong; cryptographic of the 1st Infantry Regiment under Comrade Warrant Officer Dzuong Tan Hoa; cryptographic of the 2nd Infantry Regiment under Comrade Warrant Officer Ngo Xuan Tu; and cryptographic of Group 80[28] under Comrade Warrant Officer Tran Van Tuoi, etc.

From its establishment until May 1962, the Region Cryptographic Bureau had a Party team colocated with the cell of the Operations Bureau. In June 1962, the first [Party] cell in the [Cryptographic] Bureau was established.

The Bureau table of organization comprised sections for encrypting and decrypting messages, and a technique research section.

Because of having to worry about transportation for the necessities of life and transportation of cryptographic materials and professional means for the CPs ahead and behind and a number of units, in 1965 administrative sections [hanh chinh quan tri] were established. Means of transportation were mainly pack bicycles, with two Honda 90s and an electrical generator. Mobile organization sections were also established by the Bureau.

Three cryptographic comrades from MR 8 arrived at the Military Affairs Section of the Central Nam Bo region on 8 October 1961 and immediately set to work developing a liaison net with the General Staff, the Region, and with the 261st Battalion.

At this point MR 8 cryptographic was in contact with HQ and Region using type KTB4, and in contact with the 261st Bn by KTA. From the end of 1961 to March 1962, Comrades Minh and Hoa, cryppies of the 261st Bn, went to study KTB4 and returned to initiate contact with the 261st Bn by KTB4.

In the provinces, at the start, military cryptographic and Provincial Committee cryptographic were colocated [cung chung mot dau moi]. In 1965, they were split into Provincial Committee cryptographic and Provincial Unit cryptographic. Comrade Xe was in charge of the Ben Tre Provincial Unit cryptographic section,Comrade Linh in My Tho, Comrade Phu in Long An, Comrade Xuan in Kien Tuong, Comrade Nhan in An Giang, etc. Along with the nets and systems, such as MR combat commands [doc chien] (combat command 1, combat command 2); four directly subordinate battalions, 261, 263, 265, 267; the supply terminals [ben hang]:Region terminal, MR terminal, Rung Sat terminal; the schools: MR military administration, artillery school,mail units – a total of twenty-nine points.

****

In MR 9, at the beginning the cryptonet comprised these points: Region [Party] Committee, Phu Loi 1, Phu Loi 2, Rach Gia, Can Tho. In May 1963 the MR's 2nd Main Force Regiment was formed, along with the cryptographic element of the regiment, after which cryptographic was organized in the newly formed regiments and provincial units of the MR.[29]

In August 1963 we opened the Dam Dzoi campaign in Ca Mau province – the comrade chief of the MR 9 Military Affairs Section came right to the front cryptographic [element] to work in time to grasp the situation and command the two Region Main Force regiments and the Ca Mau province regional force battalion, wiping out two military sub-zones, Dam Dzoi and Cai Nuoc, in one night. The following day the units ambushed and defeated a battalion arriving as relief, and shot down many helicopters. The people of Cai Nuoc rose up and destroyed more than a hundred strategic hamlets. Comrade Bay Ngoc, the cryptographer, went to serve in this operation, and, before he was sacrificed, had told his teammate how to accomplish the mission and ensured the safety of the cryptographic material. The comrade chief of the Military Affairs Section of Region 9 commended "Cryptographic's speedy performance, a timely factor in securing victory in this action."

****

Group Phuong Dong 2, going to MR 5, also had an element of cryptographic cadre and personnel, prepared to go down and set up a cryptographic organization in MR 5, MR 6, and for Military Intelligence. Along the line of march, keeping in contact was very difficult, sometimes on the march, sometimes ensuring cryptographic liaison with HQ, with Region [Party] Committee, and with Cde Vo Bam's station (Group 559),with the three MR battalions, and internal elements of the operation – after more than two months they arrived at the regional base of Region 5 [Party] Committee.

In July 1961, the Region 5 Cryptographic Section was formed and made directly subordinate to the staff elements of MR HQ. At the start, MR 5 Cryptographic Section comprised five comrades, with comrade Nguyen Van Long as section chief. At the end of

November of that year the section was augmented by four comrades just returned from a Region [Party] Committee refresher course.

The initial cryptographic liaison system comprised a cryptonet with HQ, with Region, with the sector Party committee, with a few of the main force battalions of the MR, and, afterward, with the provincial units. Cadre and personnel overcame many adversities and difficulties initially and [with] the number of people to work, with respect to professional means, ensuring service to guidance in building the armed forces and building the revolutionary movement, ensuring service to small battles while at the same time increasing support to production of foodstuff, building the basis for expansion for MR 5 cryptographic from that point.

The following years (1962–1964) were years of even greater revolutionary struggle by the people, during which the armed forces of the region expanded, and a system of cryptographic organization was established broadly with the directly subordinate cryptographic organizations: cryptographic of the main force units, three regiments, one infantry battalion, two sapper battalions, a number of branch units, and the provincial units.

From the standpoint of technique, the cryptographic units in the MR essentially used KTB 4. Annually the MR cryptographic section organized the receipt of tons of cryptographic materials and means from the North brought down via the western Truong Son [Ho Chi Minh Trail] line of communication, having to man-pack for two to three months to reach the MR rear area, but still not having enough to use. The Technique Research element of the MR Cryptographic Section had to produce code chart systems and handbook systems themselves; had to duplicate cryptographic key to arrange for outstations.

The Cryptographic Bureau was very zealous in assisting Sector 5 with people and means, but the augmenting groups of cryptographic cadre and personnel were inadequate for expanding in accordance with the expansion requirements of the armed forces. The Cryptographic Section organized training and development so as to have people to work.

At the end of 1964 and beginning of 1965, the revolutionary movement in the Sector had become powerful. Cryptographic forces in the main force troop units of the MR and provincial units ensured the secrecy of command in battles, eliminating the mass of small enemy posts in the mountainous jungle and the border plains, helping the masses destroy the "strategic hamlets."

In Region 6 the cryptographic organization comprised the Region Cryptographic Section under Comrade Zuong Tan Dong and the directly subordinate cryptographic organizations.

The Region Military Intelligence [tinh bao quan su] Cryptographic Section with the cryptographic organizations directly subordinate, the sectors, groups [cum] – Cde Truong Tan Them was the section chief initially.

****

Alternately building and performing their task in conditions of endless difficulty and hardship, the cryptographic organizations of the Liberation Army served guidance and command in the political struggle, opposing the strategic hamlet posts established by the people, and served guidance in the victorious destruction of strategic hamlets. Especially in the Binh Gia campaign (late 1964), the Cryptographic Bureau of Region HQ [Bo chi huy Mien], along with cryptographic of the 9th Division and other units, organized and did a good job of carrying out the "campaign cryptographic task," drawing a number of experiences with respect to organization, technique, professional staff [work], etc. The Southern cryptographic organizations did a good job of serving leadership, guidance, and command in defeating the tactics of "armored mobility" and "helicopter mobility" as in the dry season plans of the American imperialists, such as the Staley-Taylor and Johnson-McNamara plans. The cryptographic organizations of the Liberation Army took part in defeating the American imperialists' special warfare strategy. Through combat service, cryptographic cadre and personnel displayed boundless loyalty toward the Party and the people. There were many examples of dauntless sacrifice to avoid leaving cryptographic technique and secret contents of the Party and army to fall into enemy hands. A number of units and individuals were awarded decorations – many cadre and personnel were presented with the appellation, "Ap Bac Warrior," "Valiant Soldier of Victory," and "Emulation Warrior."*

COMPLETING THE CHANGEOVER TO THE USE OF TECHNIQUE KTB4 AND PREPARING FOR A NEW DEVELOPMENT OF CRYPTOGRAPHIC TECHNIQUE IN THE ARMY

In September 1959 the Central Cryptographic Section and the Cryptographic Directorate of the General Staff organized a conference of cryptographic cadre nationwide and army wide, in order to thoroughly grasp "A platform for the development of technique." In this conference, an ideological struggle concerning technique brought good results. Conservative attitudes toward technique KTA, along with subjective attitudes satisfied with the strong points of technique KTB rejected the strong points and contributions of technique KTA and severely criticized both. In order to counter espionage activity, technical intelligence, and the collection of cryptanalytic information at the high technical level of the enemy, the conference decided on the line for expanding the technique of the army cryptographic branch, namely:

---

*Ap Bac, in Dinh Tuong Province, thirty-five miles southwest of Saigon, was the scene of a 2 January 1963 action in which a "VC" battalion, outnumbered four to one by forces of the Republic of Viet Nam, supported by armor, artillery, and helicopters, trounced the republican forces and escaped, killing three American advisors and shooting down five helicopters. – Tr./Ed.

"Never cease to raise the level of the KTB system in all aspects: it is essential in using the two-part codebook system, to take compound words as the basis; superencipher with objectively randomized numbers; apply a method of constructing and making them more suitable for each condition and circumstance of assignment; implement the principles for the use of bilateral [or "link," tay doi] cryptographic systems (own sending/own receiving, own sending/general receiving); restrict the use of general systems (one general codebook, own random numbers); on the principle of technique KTB, prepare the basis for gradual mechanization of encryption and decryption." [Own, or personal [rieng], in the sense of being limited to one user or link, point-to-point, as opposed to a general, or common [chung], "circular" system shared with others. Tr./Ed.]

Implementing the above line, cryptographic organizations at the various levels throughout the army zealously overcame difficulties and carried out the changeover to using technique KTB in the North in 1960 and in the South in 1962. The types of KTB systems that were researched, produced, and used during this period were the "two-part codebook ["dictionary"], superenciphered, random numbers" code (some nets continued to use codebooks that were not two-part); the "handbook, superenciphered, random numbers" code; the "chart, two-part, mixed key" code; which was very compact, used in special assignments; and the "mnemonic" code, used as a back-up to the main code in those circumstances in which the [main] code could not be carried on a special assignment.

The contents of the codebook were compiled more abundantly, gradually more suitable to the content of leadership, direction, and command in each unit and in each time setting. Methods of arranging encrypted elements and plain elements made it convenient to do the work of encrypting and decrypting quickly and accurately.

Random key was developed and produced according to a production formula and regularly inspected and rated with respect to quality, so that the degree of protecting secret information was more reliable.

Principles for the use of codes were closely determined and implemented more strictly. There was implemented a step of personal encryption, personal decryption and "restrict the sphere of use of general codes."

In the research task, we began to delve seriously into the basic problems of technique, such as the frequency of plain elements, principles and formulas for the construction of random key, etc.

Based upon the mission situation as handed on from above, in 1964 the army cryptographic branch set the technique task line for 1964-1970 as: "Do not cease to raise the level of the 'superenciphered dictionary code with key randomized' type of system as circumstances permit, in order to research and produce systems with a higher level of protecting secrecy and use consistent with the requirements of the mission essentials and task conditions, changing from digital systems to literal systems; implementing through solid, urgent steps the use of bilateral systems and restricting the sphere of use of general systems; beginning to make basic preparation to create conditions for proceeding to mechanization of the process of encryption and decryption; aiming to ensure the trustworthy secrecy, speed, and accuracy of every secret matter of the Party, nation, and army through cryptographic systems, resisting all of the enemy's schemes to collect information through cryptanalysis, which is the object of the American imperialists and [their] lackey gang."[30]

Implementing the above line, in the Air Defense-Air Force service, from the level of the service down to division, they set up a register to review codes. Through the practice of counting the frequency in the number of plain elements in the cryptographic systems, the results were that plain elements were used in a number of concrete systems: The general code of the services, 66 percent. The radar code, 68.50 percent. The air defense code, 67percent.

The air force uses many specialized terms, among which are many different types of foreign words, such as French, Chinese, English, and Russian, but at this time many terms had not been transcribed [into phonetic Vietnamese]. If one wished to encrypt and decrypt quickly and accurately, then one of the factors for technique was the construction of a special cryptographic system for the air force. With the direct help of the Cryptographic Directorate, and based on the real-world task of encrypting and decrypting, a set of cryptographic materials carrying the special characteristics of the air force was produced and relatively well done, laying the basis for future systems to be even better, consistent with the mission requirements of the air force.

Carrying out the direction of the branch's technique task, in 1964 the Technique Research Bureau of the Cryptographic Directorate expanded its research, development and production of "superenciphered, random letters dictionary" code [luat "tu dien ma kep chu loan"], having the effect of raising the level of security and reducing by one quarter the number of groups in the encrypted message. But there were many difficulties that could not be overcome in a short time, so the use of "superenciphered random letters dictionary" codes was not implemented.

Also nothing could be done about the matter of "basic preparation to develop conditions for progressing to the mechanization of encryption and decryption," for there were difficulties, both objective and subjective, among which the main bad points were that we did not yet have a regulation on training, lesson plans, the instructor corps, essential cadre, and other aspects not yet accomplished for the upper echelons and concerned organizations to perceive clearly the mission substance of the cryptographic technical task and the requirements for advancing on the path of upward development for the branch.

However, by 1964 the army cryptographic branch had completed the changeover to technique KTB4 throughout the army, and raised a notch the level and quality of cryptographic materials. At the same time, some of the preparation for taking a new step in the army's cryptographic technique was made.

****

On the basis of a thorough grasp of the situation and the new mission and the political line and the military line of the Party, the army cryptographic branch obtained training for the central task of building the branch in terms of politics, frame of mind, organization, technique, and professionalism. The third army-wide cryptographic cadre conference mobilized the "three goods" emulation movement – "good study, good task, good implementation of task regulations" – and set awards for units with the highest achievements.

In the mobilization to build the army and progress towards regularization and modernization, these emulation movements in the study of politics, military matters, technical culture, and professionalism brought the entire branch to a fever pitch. In organizations that were concentrated or in odd teams on assignments, from the mainland out to the distant islands, in secluded border guard posts or on ships operating on the high seas, in the South, inside the North, and in the friendly nation of Laos – everywhere – cadre and personnel were caught up in study.

The victory of the 1961 winter-spring political reeducation program built the strength to step up the study emulation movement, to get into depth and to receive better results every day. The cryptographic organizations carried out emulation with positive action slogans, such as "Resolve to Emulate for the Highest Prize in the Task and Training in order to Achieve the Challenge Banner from the Central Cryptographic Section," or "Study Days, Study Nights, Surpass the Norm, Strive for First Place". . . Many units implemented "the three builds" – "advance in thinking, advance in work-style, advance in technique" – and "the three counters" – "counter job dissatisfaction, counter inaccuracy, counter conservatism."

The task regulations were an essential component of the annual study program. Besides study and implementation in the branch, the task regulations still had to be grasped fully and implemented in the command organizations and in other organizations that used cryptography for secret messages. As a result, the implementation of fixed principles from the outset was more serious and strict, producing more practical results throughout the military.

The training to elevate the level of technique in use was implemented with many lively forms. Each cadre, each of the personnel, individually trained in basic technical subjects, practicing encrypting and decrypting. Cryptographic organizations went on to organize reviews of work style, reviews of technique, exchange of experiences between their unit and different units, organized inspections of technique, and held technique teach-ins. Study became a routine, with a concrete quota plan, guidelines, and suitable methods. In 1962 the army cryptographic branch organized an army-wide cryptographic techniques competition and exhibition to examine, evaluate and report on the results of training. The comrade commanders of higher echelons visited the occasion and were happy to see the progress of the army cryptographic branch. New records of productivity and accuracy were set, from 400-500 groups/hour before to 600-700 gps/hr. The prize for high productivity in encrypting (720 gps/hr) went to cryptographic of the Air Defense – Air Force, while the high productivity prize for decrypting (1,124 gps/hr) went to Navy cryptographic. The level of accuracy rose from 99.50 to 100 percent.

Professional training became more realistic, more real-world. The cryptographic organizations organized task simulations as part of tactical and campaign simulations by the staff organizations at the various levels. The cryptographic organizations, along with the communications organizations, practiced simulations in the field, the process of taking in and sending on, encrypting and decrypting, receiving and transmitting messages. In the parachute troops, cryptographic personnel trained to parachute into a hot drop zone and immediately set to work. By many other positive measures, the army cryptographic branch raised the level of organization for the cryptographic task of ensuring secrecy by cryptographic techniques under combat conditions.

In the organization of training, the army cryptographic branch attached special attention to refresher training to raise the level of capability for cadre at the various levels by means of phased training and continuing education in technique. The comrade commanders of the Cryptographic Directorate got involved in writing material and teaching, [producing] "Some Problems in Professional Instruction" (1960), "The Cryptographic Task in Modern Combat" (1962), etc. The cryptographic organizations of the MRs, services, and divisions also summed up experience for cadre refresher.

The task of training at the school, with respect to syllabus, pedagogy, and training, never ceased to be raised in quality of training, so that students leaving the school to return to their units after a short period of refresher would be able to perform in the real world. This was a very great effort on the part of the school, and it was praised by the entire branch and commended by higher echelons.

The cryptographic organizations in the South cleverly combined on-the-job training with the real-world task, taking advantage of time for training, aiming at having cadre and personnel adjust to each circumstances of work or combat. Cadre in the basic units became the nucleus of the "two goods" emulation campaign – "good work, good study." Along with refresher work to raise the level of old cadre and personnel, the military region cryptographic organizations organized training at the local level for many new personnel. From 1962 to 1964, 437 comrades were trained in theaters B1 and B2.

"A worry-free ideology of long service in the branch had clearly determined that there would be more basics and essentials than before. The comrades assigned in places of much difficulty, hardships, and sacrifice have held on to their positions, broadening the traditions of the branch in a most satisfactory way. Comrades perform the mission in places lacking accommodations with respect to material and intellect, yet continue diligently in the task, performing their mission well. Many comrades sought out assignments in places with the greatest difficulties and hardships."[31]

In the annual accounting of mission accomplishment and training, many cryptographic organizations received the "Best Training and Performing Unit" banner, such as the Cryptographic Branch of the Naval Arm, the MR 4 Cryptographic Bureau, the

Right Bank MR Cryptographic Bureau, the Cryptographic Sections of the304th and 330th divisions, and the Cryptographic Section of Group 959.

DUTY IN THE LAOTIAN THEATER

From the end of 1954, the American imperialists replaced the French colonialists and instituted a neocolonialist policy in Laos. America manipulated its Lao puppets into attacking the two provinces of Sam Neua and Phong Saly, liberated areas of the Laotian revolution. Under the leadership of the Lao Patriotic Front, the Lao army and people struck back decisively, inflicting heavy losses on the enemy and forcing them to sign the Vientiane agreement, establishing the first national unity government in Laos, including delegates from the Lao Patriotic Front. Before the Lao patriotic forces could occupy some important seats in the unity government and while the prestige of the Lao Patriotic Front was at its highest, the Americans and their henchmen pulled new tricks of sabotage. In May 1959 they seized as prisoner Prince Souphanouvong, and a number of other leadership cadre of the Lao revolution. At the same time, they surrounded and issued orders to disarm the two battalions, the 1st and 2nd, of the Lao People's Liberation Army, intending to eradicate the Lao revolution. After being surrounded for a week, the 2nd Battalion and an element of the 1st Pathet Lao Battalion valiantly and cleverly broke through the enemy encirclement and got back to the secure base area with the aid and sympathy of the people of the regions. The PAVN cryptographic organizations did a good job of serving command in the organization of the reception of the 2nd Battalion in its victorious escape.

In July 1959, PAVN cryptographic organizations served the command of Group 800 of the Vietnamese Volunteer Army in opening action on the Vietnamese-Laotian border, wiping out the Sam To and Nong Het forts,and helping our friends expand their territory and their forces.

On 23 May 1960, Prince Souphanouvong and other Laotian comrades successfully escaped jail: The Encrypting-Decrypting Bureau of the Directorate of Cryptography, the Cryptographic Section of Group 959, and the cryptographic organizations of the Vietnamese Volunteer Army did a good job serving command and control in meeting Chairman Souphanouvong and the leadership comrades in their return to the safe base region.

The Lao Patriotic Front expanded. While the cryptographic organizations of the Vietnamese Volunteer Army served the command in victorious combat, the air force cryptographic organization arranged cryptographic teams[32] assigned to the Plaine des Jarres, Sam Neua, Vang Vieng, and Tchephone airfields, serving direction, command, transport and the paradropping of supplies, weapons, and equipment for the friends and ourselves. In 1960-61, the cryptographic organizations of the 959th Group, MR Northwest, the Air Force Directorate, and other units completed the mission of ensuring command secrecy in the Plaine des Jarres-Xieng Khoang Campaign. In 1962, the organizations of the General Staff Cryptographic Directorate, the cryptographic organizations of MR Northwest, the 959th Group, the Air Force Directorate and other units participated in the major victory of the Nam Tha campaign.

In these campaigns, the cryptographic organizations of the PAVN all accomplished the mission. Many units and individuals did so in an outstanding manner. Cde Nguyen Ngoc Khue bravely gave his life while performing his mission at Salaphukum in April 1961, receiving posthumously the Order of Military Merit. Cdes Hoang Tu and Dao Trong Luat, cadre of the General Staff Cryptographic Directorate, fulfilled the mission and returned after a special assignment trip into the enemy area of Laos and were also awarded the Military Merit medal. Service achievements of the Vietnamese Army cryptographic organizations were part of the great victory of the Lao army and people. In June 1962 the Geneva accords on Laos were signed, recognizing the legal status of the Lao Patriotic Front and undertaking to respect the independence, neutrality, utility, and territorial integrity of Laos, and establishing a tripartite government of national unity.

During this time, the 959th's cryptographic forces had 130 comrades (including cryptographic specialists helping the friends at Central and an element helping the friends at the Xuan Thanh cultural school). Alternately working and building, in conditions of endless difficulty and hardship, the 959th Group cryptographic organization accomplished their mission well. Implementing the 1962 Geneva accords on Laos, the cryptographic organization belonging to the 959th Group HQ returned home. Almost all cadre and personnel were appointed to assignments in the military regions, services and branches.

The accords had barely been signed before the American imperialists and their lackies plotted to subvert the 1962 Geneva Accords, nurturing the reactionary forces of Vang Pao and mobilizing the army of Thailand to cross over and, in succession, to open large operations aimed at the destruction of the Laotian patriotic forces.

As requested by the friends, in November 1963 the system of cryptographic organization in which the Vietnamese Volunteer Army was directly active was returned. The cryptographic organization of the 959th Group was divided into two elements and went in two directions – one over to Sam Neua, one over to the Plaine des Jarres and Xieng Khouang.

The cryptographic organizations of the 923rd Battalion of MR 3, the 924th and 927th Battalions of MR 4 (each battalion having two cryptographic teams), along with other units, crossed over to perform the mission in Laos. The cryptographic organization of the 463rd Subgroup [phan doan] was established and active in the Plaine des Jarres. The Cryptographic Section of the 959th Group alternated between performing its mission of ensuring command secrecy and helping the friends' cryptographic branch straighten out its organization, bring up to strength the ranks of their cadre and personnel, and prepare technique and equipment, so as to be ready to meet the growing mission requirements of the Lao revolution.

In 1964, the cryptographic organization of MR Northwest combined with the cryptographic organization of the 959th Group and other units and did a good job of implementing the cryptographic assignment of ensuring command secrecy for the Plaine des Jarres-Xieng Khouang campaign. The Cryptographic Bureau of MR Northwest, the cryptographic sections of the 316th Division and 335th Brigade, along with the Cryptographic Section of the 959th Group – all fulfilled the mission on the soil of our friends in an outstanding manner.

Notes

1. During the resistance against France, there were over 30 points in liaison with HQ. In 1955-56 there were about 50.

2. In 1955, 10 cadre-in-charge and 108 cadre and personnel; the first six months of 1956, alone, 41 cadre and personnel.

3. Cryptographic Section of the 332nd Division under Cde Nguyen Trung Nghi. Cryptographic Section of the 330th Division under Cde Nguyen Cong Tru. Cryptographic Section of the 335th Division under Cde Nguyen Hoang.

4. Comprising 26 Cryptographic Sections in these units: General Directorate of Supply, the Central Joint Cease Fire [Commission], the Directorate of Civil Aviation, the Directorate of Intelligence [Tinh bao], the Coastal Defense Directorate, MRs Viet Bac, Northwest, Left and Right Banks, 4, Northeast, 367th Division (Air Defense), 351st Division (Artillery), Group 100, divisions 304, 305, 308, 312, 316, 320, 324, 325, 330, 332, 335, and 338. At this time the engineer cryptographic teams were subordinate to the General Staff Cryptographic Bureau.

5. Extract from a speech by Cde Hoang Van Thai at the eleventh cryptographic conference.

6. Signed by Cde Nguyen Dzuy Trinh, representing the Secretariat.

7. Resolution No. 59-NQ-TW dated 20 November 1958, signed by Cde Nguyen Dzuy Trinh representing the Secretariat.

8. The predecessor was the Cryptographic Section of the Coastal Defense Directorate, which, from May 1955, had been under Cde Le Dinh Lien.

9. Patrol boat group with Cde Cao Van Duc in charge of cryptography. A 79-ton ship with Cde Le Van Dien in charge of cryptography.

10. Such as Lai Chau, Dien Bien, Moc Chau, Song Ma, Quy Chau and a number of border defense posts--Muong Nhe, Muong Loi, Leng Xu Xin, Sop Cop, Na Dit, Keng Du, Vinh Linh, and Huong Lap. [See the supplement to this edition -- Tr./Ed.]

11. This group finished its duties and was dissolved.

12. With three subordinate units.

13. Namely Huoi Phong, Muong Xen, Hon Co, Na Luong, and MR 4 turned over 5 stations to Armed Public Security.

14. Bac Ninh Provincial Unit and two stations serving timber exploitation. MR Left Bank turned over stations at Cat Ba, Thai Ninh, and Dinh Lap and the maritime unit to the Navy.

15. And withdrew cryptographic in five units, the 154th and 169th Regiments, the antiaircraft battalion, and the Son Tay and Ha Dong provincial units.

16. Namely the 905th and 907th Battalions, Nga Truong, Than Uyen, and 168th Artillery Regiment, and handed over four stations – the 953rd, 955th, 957th, and 959th battalions – to Armed Public Security.

17. Lai Chau Airfield.

18. Cryptographic in three battalions – the 54th, 55th, and 3rd – were withdrawn at the beginning of the year and reorganized in December.

19. One accelerated class comprised fifty-two comrades, one had eighty-one.

20. Of the grand total of 291,152 official messages of Cryptographic, nationwide.

21. Extract from document of the eleventh army-wide cryptographic cadre conference, 1954.

22. From the proceedings of the eleventh army-wide conference of cryptographic cadre, 1954.

23. From a speech by Cde Hoang Van Thai at the 1956 army-wide conference of cryptographic cadre.

24. Patrol sector 2 at the Song Gianh, Cde Pham Minh, Chief of the Cryptographic Section.

25. Type HTA.

26. Group 759 changed designators and became 125 of the Navy.

27. Afterward, M8, Bureau 8.

28. Predecessor of the Region's artillery division

29. Namely Can Tho, Soc Trang, Vinh Long, Ca Mau, Tra Vinh, and Rach Gia (each province still having a cryptographic team in the regional battalion).

30. "Directions for the Technical Task in 1964," on file at the Cryptographic Directorate of the General Staff.

31. From the report recapitulating the training task for 1961.

32. Comrade cryptographic personnel Xuyen, Muc, Khue, and Dinh.

Chapter Six

Ensuring Leadership and Command Service in Beating the Escalating War of Destruction of the American Aggressors in the North and the Violent Local War in the South (1965–1968)

CONTINUING TO BUILD AND TO EXPAND FORCES WHILE SERVING TO DEFEAT IMPERIALIST AMERICAN DESTRUCTIVE WAR IN THE NORTH

Defeated in carrying out their neocolonialist policy by "special warfare," the American imperialists changed over to "limited war" in the South and stirred up a war of destruction in the North through their air force. They massed and brought over tens of thousands of American troops and those of their satellites into the South, and speedily increased the troop strength of their puppets. They mobilized thousands of modern aircraft along with a large part of the forces in their Seventh Fleet to strike the North.

Confronted by the frenzied plots of the American imperialists, on 27 March 1964, at the capital in Hanoi, President Ho Chi Minh convened a special political conference to raise the resolve of the people, namely their determination to solidly guard the North's socialism, while at the same time lending aid to our blood brothers in the South, and uniting our nation.

Imbued with the resolute ideology of the Central Party Politburo and the Central Military Committee, the policy line was to build peoples' armed forces in the two areas, South and North, that were strong both in numbers and in quality.

The system of army cryptographic organization also expanded wider and deeper with every passing day, in the command system of the armed forces of our people.

In 1965, the army cryptographic net increased speedily and in complexity (one net had 37 units with cryptographic organizations, but the number of points rose to 111 – internal contact, skip echelon, and combat coordination). Compared with 1964, the army-wide cryptographic net increased by 203.54 percent.

In 1966 the army cryptographic branch had 3,205 cadre and personnel, yet still was unable to meet mission requirements completely and in timely fashion.[1]

Cryptographic organizations lacked people at all levels; almost all basic units and many regiments had only one cryptographic comrade performing the mission. The daily labor intensity was high. In the Cryptographic Bureau of MR 3, compared with 1964, 1965 saw a 138 percent increase in the number of liaison nets and a 76 percent increase in troop strength, but the volume of messages encrypted and decrypted doubled.

The Central Cryptographic Section and the Cryptographic Directorate proposed to the Secretariat and the Central Military Committee the direct selection of good party members, both male and female, in production bases, business organizations, and work and agricultural sites, to be trained as cryptographic personnel.

The General Staff gave enrollment quotas to the MRs, services, branches, organizations, and units. The comrade commanders of staff organizations at the various levels concerned themselves with guidance and inspection in speeding up the accomplishment of plans for assigned quotas. The cryptographic organizations coordinated closely with military personnel organizations and guard organizations to accomplish the quota plan.

During this time the Army Cryptographic School again had to evacuate in order to ensure security; thus it met many difficulties in organizing living conditions and in teaching and study. Many classes had to be expedited to meet mission requirements. Although the curriculum and duration had to be reduced, still the school was able to ensure quality training.

The school board paid particular attention to fostering political ideology and professional technique; at the same time they had to train and foster physical strength to give each student sufficient health to take to the field and perform their mission in distant theaters. The cadre administering the students and the instructors made many efforts to improve the content and methods of teaching, reserving much time for technique practice and field exercises. Students were fired up to emulate in study, including days off and hours off reviewing training, in order to firmly grasp professional technique with the impatient wish to soon go and receive their mission, mainly to be able to go down South to fight and liberate the homeland. Cadre and personnel serving at the school also made efforts to emulate in ensuring good life style, spirit, and material, participating in the performance of training and giving refreshers to cadre and personnel faced with the new requirements.

Cryptographic nets expanded quite rapidly – the task of research and production of the technical means was stretched to the fullest. Measuring wits with the enemy on the battle front of cryptography and cryptanalysis a form of combat that is extremely arduous and decisive. With the slogan "sweat on the desk reduces blood on the battlefield" in their silent work, the comrade cryptographic cadre and personnel did the work of two, because the South was part and parcel [of the nation].

Based on the plan for code research, the technical research organization achieved 110.58 percent of the number of thick codes, 100.46 percent of the number of average codes, and 141.66 percent of the number of chart systems. (During this time, the Techniques Bureau of the Directorate was augmented by seventeen cadre and the printing plant by forty-seven workers.)[2] The regular and first priority job was to compile codes and key quickly to serve for encrypting and decrypting, but also not to lightly dismiss projects of a basic nature. In 1966 the Techniques Bureau researched and fully worked out formulae for creating cryptographic key, reviewed and selected 302,307 plain elements of code, completed a number of types of code that accorded with public documents, to be used for a number of special nets and intelligence [tinh bao] nets, and executed a recapitulation of chart-codes.

Implementing the General Staff directive concerning the task of ensuring command secrecy over ultrahigh frequency communications, starting in 1966 cryptographic of the Air Defense-Air Force Service and Cryptographic of the Naval Service were given the additional mission of compiling and producing operational codes [mat ngu] to serve combat command in the sphere of their services. This was a brand new task and extremely complicated, for a cryptographic bureau of that period. Message usage was wide, the volume large. The objects of use were command cadre, the pilots of the air force, the skippers of the navy,[3] watch officers, and cadre of different specialities. The requirements of structure had to be simple and easy of use so that command could be rapid and relatively secret. This posed a difficult dilemma to resolve – on the one hand the matter of secrecy, on the other, convenience in use to facilitate speed. The Techniques Research Bureau of the Cryptographic Directorate assisted the two cryptographic bureaus of Air Defense-Air Force and Navy in overcoming this obstacle. They alternately worked and took advantage of the participation of the organs of command, operations, and communications to show the way; additionally, they went down to the units, seeking to understand terms used in oral commands, reporting, means of communicating when exercising command; seeking to understand the tactics and techniques of the branches, their ordnance and equipment, the equipment in their command posts, etc.

The Air Defense-Air Force Cryptographic Bureau made up many types of opcodes to serve the various services – radar, AA, rocket, air force, and other organizations.[4]

The Navy Cryptographic Bureau researched the various types of opcodes, HTA, HTB, HTF[5] and organized guidance for directly subordinate cadre to use. Annual production was a thousand charts, a thousand books of various types, from simple use to ciphers with changing keys.

That way they did their part in ensuring command secrecy when conversing by shortwave communications, reducing the volume of messages that ordinarily did not have to go through cryptographic [handling].

In May 1966, the Cryptographic Section of Engineer HQ was established, with nine directly subordinate cryptographic organizations and comrade Truong Cong Nghi as section chief. From that, the engineers' cryptography developed swiftly, serving leadership and command of the engineer troops in building and fighting in the two parts of the homeland.

In May 1966, the Navy's Water Sapper [Accepting the quaint old military term, which has become the standard American rendering of the Vietnamese dac cong, from cong tac dac biet, special task or special assignment: these elite specialists are comparable to commandos, rangers, SEAL teams, etc., in the West. – Tr./Ed.] Group 126 was established, with comrade Nguyen Chuong Lien as chief of the group's cryptographic section. The group organized a forward element called Group 1A, its cryptographic organization chief being Cde Ngo Hai Thuyen, serving the advance CP. Water sapper activity to the South increased greatly. The 12th Company, 4th Battalion, of the group trained and professionally upgraded cryptographic comrade augmentees at the naval bases and dockings: twelve teams of cryptographers, comprising twenty-four comrades, alternating between professional study and lying beneath the surface, swimming with a breathing tube, using sapper tactics, prepared to receive orders to set out right to the units serving command, participating in combat until the South was liberated – such comrades as Hoi, Chau, Hung, Bon, Nhung, Dang, etc. There were comrades, such as Cde Ngo Van Dang who bravely gave their lives when one of our boats was discovered by the enemy and they had to destroy it. Also, from the time of its establishment until the signing of the Paris Accords on Viet Nam in 1973, Navy Group 1A had sunk hundreds of American warships on the rivers of Cua Viet, Dong Ha, Dzuy Phien, Dai Do, Xuan Khanh, and Quang Tri. By day and night, cryptographic comrades of Group 126 served command, taking much pride in their unit, but also doing their utmost in labor to have a part in the glorious saga of the group--twice cited for heroism among the armed forces in the long years of arduous and violent combat.

From 1965 to 1968, Navy cryptographic served competently and heroically in the battles of the Vietnamese People's Navy, both North and South, making many achievements.

The Cryptographic Bureau of the General Directorate of Rear Services [GDRS] was established 16 December 1966, directly subordinate to the staff of GDRS, and with thirty-two liaison nets at the beginning. A number of cryptographic organizations from rear services units in Group 559, military relay stations, various odd radio stations in Na Meo and Muong Sen, depots 710 and 486, previously subordinate directly to the General Staff Cryptographic Directorate, were turned over to the GDRS Cryptographic Bureau.

The cryptographic forces of the Ho Chi Minh trail [Truong Son] troops grew enormously, comprising three divisional cryptographic sections, forty-six regimental cryptographic subsections, and military relay stations. In September 1967, the Cryptographic *Section* of Group 559 became the Cryptographic *Bureau* of HQ [Bo tu lenh], Group 559.

The cryptographic organization of the Sapper branch took shape on 19 March 1967, with comrade Nguyen Si Chuong as section chief. At the beginning there were only cryptographic organizations in the 24thBrigade, the 246th Regiment, and the 126th Regiment; afterward, the organization of cryptography quickly spread in the units of elite special troops that wormed their way deep into the enemy's heart to strike air bases, POL depots, ports, etc.

In midyear 1967, the Western Area Task Committee (CP38) was formed on the basis of merging CP31 and the 959th Group, the cryptographic section of CP38 having, at that time, thirty-six people, covering liaison for thirteen units.

By July 1967, the system of army cryptographic organization in the North had 1,106 liaison points. The volume of outgoing and incoming secret messages had increased manyfold, because the cryptographic warriors were present in almost all units from the mainland to distant islands, ports, naval bases, relay stations, riverside depots, main points in the lines of communications, the military groups strengthening the South, the mobile command stations [engaged in] capturing enemy flight crews and destroying delayed-explosion bombs to unclog roads and vehicles, and military intelligence [quan bao] teams.

In 1966 army-wide encrypting and decrypting elements encrypted and decrypted 1,982,225 secret messages, a twofold increase over 1965. In MR 4 in 1965 there were 274,708 secret messages, as compared to 191,681 in 1964. In the first eight days at the start of the war of destruction, the Navy's Cryptographic Bureau encrypted and decrypted 2,756 messages. In 1965, the Message Encryption-Decryption Bureau of the Cryptographic Directorate had to take care of 105,845 cryptograms comprising 8,933,449 groups. Urgent messages went and came flooding, to the point that they had to be handled on time and completed, whether day or night. But thoroughly permeated with the strategic determination of the Party, as the victory messages came flying back with enthusiasm, the young cryptographic boys and girls worked without tiring. Responding to strictly time-sensitive requirements, there were messages that had to be figured in minutes – many in which, in the space of an hour, it would have been over with. Communiques concerning activity of the enemy's B52 planes, communiques of naval gunfire shelling the mainland, communiques of the objective and time the enemy would strike; the posting of combat forces, adjustments in vehicle formations, changes in troop stationing; command of campaign transport, direction of groups on the march into the South to overcome the main points of enemy attack; command of diversionary troops, to create the sudden destruction of enemy aircraft, etc. The command contents necessitated having to ensure absolute accuracy – one incorrect letter, one digit, could cause loss of people and property. In order to act in a timely manner and ensure accuracy in message content, the cryptographic cadre and personnel and those of communications had to constantly keep up the level of their sense of mission responsibility in the matter of total message handling, in decrypting and encrypting with patience. By natural general knowledge and real world experience, the comrades discovered and rectified tens of thousands of cases of errors by fellow cryptographers, from the process of handling the encrypted signals [tin hieu ma] of radio and from people using secret messages. Thus the quality of service was raised, elating the command echelons and resulting in kudos.

In 1965, the Cryptographic Bureau of MR Northwest rectified message content accurately and promptly submitted immediate secret messages of the General Staff communicating the day and hour the enemy would strike the vicinity of the troop cantonment of MR HQ. MR HQ ordered the organizations, units, and the people to speedily evacuate the night before. At 0700 the following day, the enemy struck. We brought down an American airplane and suffered no loss of people, weapons, equipment, and material. The cryptographic cadre and personnel were commended by the MR for prompt, accurate service.

The first victories of 3 and 4 April 1965 in the airspace over Ham Rong were inscribed in resplendent golden lines in the history of our air force. That feat of arms had the labor of all collectivity supporting on the ground, including the behind-the-scenes contribution of Air Defense-Air Force Service Cryptographic.

Right from the first days of March 1965, from the [Air Defense-Air Force] HQ organizations, the Cryptographic Bureau sent along message communiques assessing the situation, estimating the enemy's schemes, reminding the units to be vigilant, to make combat preparations and to train according to plan, especially the joint operations plan for the branches.

Messages from the Staff [Bo tham muu] reminded the units to check and augment the operations plan, reminded them concerning the joint signals set up between the air force and the antiaircraft, also to have the cryptographic watch for encrypting and transmitting speedily. The political, rear services, and technical organizations of the service also sent messages of guidance to ensure the tasks, and mobilized a spirit of emulation to prepare to recognize the occasion of Uncle Ho's presenting the service with the "Resolve to Strike the Aggressor American Enemy Victoriously" challenge banner.

On the morning of 3 April, numbers of meteorological documents were encrypted and decrypted accurately by cryptographic and sent to the command post promptly to help HQ have the basis for ordering the use of the forces. The enemy's capacity for a large strike was clearly anticipated by HQ from the outset: the Party Current Affairs Committee of HQ met in extraordinary session, with political commissar Dang Tinh chairing, to analyze the situation, release decisions, and discuss means of implementing the operations plan of the service as reviewed by the standing committee of the Central Military Committee and the General Staff – if the enemy struck above the twentieth parallel, our air force was to attack. HQ organized for joint battle involving the air force and AA in the Ham Rong zone. The resolution of the Current Affairs Committee and intention of HQ became orders in effect and at once, through cryptographic organizations at the various levels encrypting, decrypting, and forwarding on to the basic units along with other means of communication liaison. At the same time, on the joint network, service cryptographic organizations swiftly encrypted to go a message announcing that our aircraft were to make combat attacks, up to MR 3, MR 4, MR Northeast, and the naval service. A portion of the message said, ". . . will have four to six of our aircraft in combat over the skies of Thanh Hoa. Request units examine their signals for distinguishing the enemy and ourselves and for aircraft recognition. . . ."

Each aspect of the preparations was carefully checked over by the CP at the end. At exactly 0847, the order for the first flight to take off on a support mission – a minute later, the attack flight sortied. The decisive, lightning-like combat took place in the space of just four minutes, and our Air Force had shot down two American planes in the skies over Do Len, Thanh Hoa, and returned to base safely. This was the historically significant first victory of the Vietnamese People's Air Force, a beautiful team effort by the air force, navy, and air defense forces.

After the victory, secret messages from HQ and the organizations came pouring down to the Red Star group, with instructions to organize and draw experience from the combat, and to prepare in every respect for the next day's strike. Tactical operations plans were supplemented and fully worked out; the task of preparation went straight into the night. Exactly as estimated from above, on 4 April 1965 all hell broke loose. From 0930 the enemy mobilized the naval bombing forces on the two ships, *Hancock* and *Coral Sea*, to penetrate and strike Ham Rong. The enemy determined to use "air power" to collapse the steel bridge spanning the Song Ma seventy miles [dam] to the south of Hanoi. After using naval forces to no effect, the Americans had to launch en masse an F105 squadron of their air force for a strike. Our air force attacked. At exactly 1020 the first flight took off, with diversion its mission; two minutes later, the attack flight received orders to leave the runway. Only a few minutes later, in the skies over Thanh Hoa, our air force broke into a sudden decisive attack. Along with the navy and air defense forces, the Red Star air force group again scored outstandingly, shooting down two American F105 "Thunderchiefs," and taking part in the general feat of arms of the military and people at Ham Rong, Thanh Hoa.

The whole nation cheered with joy when they heard of the big northern area victory on 3 and 4 April, shooting down more than thirty aircraft, beating the new escalation of the American imperialists.

Messages announcing news of the victory and congratulatory messages from the Party, from Uncle Ho, and from the High Command were swiftly passed to and fro after the victory.

The Air Force-Air Defense Cryptographic Bureau extracted experience from ensuring air force command in their first victorious strike, the air force campaign to protect Vinh city. Up to the "Route 5 Campaign" in 1967, the Air Force-Air Defense cryptographic organization did a fine job of accomplishing its mission, serving continuously through fifty-six days and nights.

Service to operations having gone this way, the transmittal and receipt of cryptographic materials, professional documents, and technical means and equipment also could not be permitted to weaken, principally vis-a-vis the distant islands, the areas of responsibility, the main points subject to continuous and violent strikes. Not fearing hardship or having to lay down their lives, cadre and personnel did a good job of accomplishing the mission of the task flights, ensuring continuous connectivity for the cryptographic techniques network in the command machinery.

Prior to the violent combat and the most important mission, the branch cadre and personnel built high resolution and [sense of] responsibility. Cde Phuc, a cryptographic warrior of air defense; Cde Lung, a sapper cryptographic warrior; Cde Nghiem, subordinate to the General Directorate of Rear Services; and many other comrades had loved ones killed and homes destroyed, yet remained with their assigned units, turning grief and vindictiveness into strength to fulfill the mission. On the island of Con Co, heroic and dedicated Cde Mai Quang Dzi, cryptographic warrior, worked day and night, ensuring the accurate sending and receiving of messages, despite the volume (some days fifty to sixty messages). There were times under the pressure of bombing, ears buzzing, head bursting, tight of breath, that the comrade continued to work. Besides accomplishing the mission responsibility of his speciality, Cde Dzi, along with hero Thai Van A, stuck close to the observation station commanding the units defeating every trick of the striking American aircraft and ships. In 1965, the Con Co cryptographic team was awarded the Military Merit decoration, second class, the comrade cryptographic personnel receiving the decoration in the third class. In MR 4, combat was furious, with many examples of heroic sacrifice by cryptographic warriors: Cde Phuc, badly wounded in both legs; Cde Mac, badly wounded in the belly and losing an eye; Cde The, taking a bomb fragment in his head; Cde Xy, burned by napalm, etc. The comrades, under pressure and in pain, stuck to their assigned positions until people arrived to treat them and replace them.

On 29 September 1966, an MR Northeast ship was attacked and sunk, and its cryptographic materials also went down. The comrade cryptographic personnel, along with some people still with the ship and regional guerrillas, resolutely groped around and brought them up.

Because the South were blood brothers, cryptographic comrades performing the mission in the North sent letters of resolve to the upper echelons, [expressing their] eagerness to go to the South to fight and work, to take part in the liberation of the nation. Comrades who received this glorious mission were elated to set out, overcoming difficulties and loss, thinking of their families in the North and the feelings and material losses of their native land due to American bombing.

In MR 3 during 1965, they prepared to support thirty-one stations and sixty-two cryptographic comrades for theaters B and C. In 1966, the Cryptographic Directorate organized for stations going to B from the General Staff and MRs 3 and 4 (consisting of forty comrades, per the tables of organization of eight regiments, a battalion of the 324th Division and MR 4 forward area) 125 sets of cryptographic systems and 13,000 sets of cryptographic organization for HQ, B5, transshipping many cryptographic organizations subordinate to the regiments, infantry battalions, artillery, units, to the tune of 475 cadre and personnel and 368 book-type systems, 23,726 sets of cryptographic key, a printing press, 25 kilograms of type and tons of equipment for Theater B.

From the end of 1967, the cryptographic organizations of the mobile reserve divisions of the High Command--the 304th, 308th, and 320th divisions – were placed on standby to prepare to enter the theater of war and were strengthened in every respect.

In the artillery branch, there were thirteen radio stations in 1967, including ten in the North. In June and September, two regiments, the 68th and 208th, went down into Theater B, the Cryptographic Section having prepared for the 68th Regiment, four stations, comprising eight cryptographers, and for the 208th Regiment, two stations with four cryptographers, etc.

In January 1968, our Central Party Executive Committee issued a resolution concerning the political mobilization of all our people to carry on to victory the mission,

"All out to beat the American aggressors." The Executive Committee and the Central Military Committee decided to open an offensive and simultaneous uprising over the entire South on Tet of 1968.

One day early in 1968, Cde Chief of the General Staff Van Tien Dzung, representing the Central Military Committee and the High Command, arrived to pay his respects and to commend the mobilization of General Staff Cryptographic Bureau cadre and personnel. The comrade thoughtfully made recommendations: the mission of the Cryptographic Directorate and the army cryptographic branch in 1968 was going to be most exacting, but most glorious. The comrades would have to concentrate their strength to the highest level in order to ensure the contents of leadership, guidance and command from the Central Military Committee, the High Command, and the various echelons of committee and office heads throughout the army – that secrecy be absolute, accuracy be the highest, and timeliness absolute. The aim of ensuring this was to secure a great victory. Cde Le Thanh Hai, Chief of the Cryptographic Directorate, on behalf of the cadre and personnel of the entire branch, pledged to the comrade chief of the General Staff to resolutely fulfill each mission that was received. Afterward, the comrade chief of the directorate sent a secret message conveying the words of congratulation and recommendation of the comrade chief of the General Staff to all cadre and personnel and mobilized the entire branch to precisely execute the recommendations of the comrade chief of the General Staff.

Serving combat both in the South and the North, each year the Army Cryptographic Directorate regularly summarized the situation, commented on the strong points and shortcomings, analyzed the causes, and derived experiences of a professional-guidance nature.

Based upon documentary summaries and recapitulations from the cryptographic organizations at the various levels, the General Staff Cryptographic Directorate compiled documents providing theory and reality in the task of professional administration and use of technique, in order to build up the cadre and personnel, documents such as "Organization and Implementation of the Cryptographic Task in Combat," "Thoroughly Grasping the Content of Compiling Code Dictionaries in order to Raise the Technical Level of Encryption and Decryption," "Checking and Predicting Errors," "Methods of Training in the Four Basic Technical Subjects and Raising the Output of Encryption-Decryption" – these were brought into play with realistic effect.

IMPLEMENTING INSTRUCTIONS CONCERNING INCREASING THE KEEPING OF SECRECY IN THE TASK OF ENSURING COMMAND SECRECY BY CRYPTOGRAPHIC TECHNIQUE VIA THE MEANS OF COMMUNICATION

From 1965, the entire nation was at war, and radio communication means became the most important means in the tasks of leadership, guidance, and command from the Party, the nation, and the army. The important economic branches, such as electricity and coal, and especially transportation lines, also used numerous stations and cryptography. In the army, as in the branches and echelons of Party and nation, covernames [mat danh] and operational codes [mat ngu] were used in conveying essential message content via shortwave or telephone communications. The system of communications expanded, greater than before; the volume of messages sent into the ether many that had not been individually reviewed and which disclosed weaknesses and shortcomings, creating favorable conditions for the enemy's information collection and cryptanalysis. Many unessential places also organized cryptography and radio; many used incorrectly the "secret and urgent" designations; many organizations copied the transmissions and kept files of secret messages in violation of principles – [there were] even many organizations that sent messages with secret contents "in the clear"; and many branches that, on their own, set up radio stations, made up their own cover terms, operational codes or simple types of cryptographic systems, compromising many secrets through their initiatives. As a result, the work of emphasizing technique for maintaining secrecy in the task of cryptographic liaison was quite essential, and had to quickly rectify the organization and use of cryptography and radio stations in order to protect the secrecy of the Party, the nation, and the army.

On 6 June 1966, the Central Party Secretariat issued "Instructions Concerning Increased Secrecy in the Task of Radio Communication-Liaison of the Party Organizations and the Nation."[6]

The instructions laid out clearly that ". . . in a situation in which the entire nation is at war, tasks expanding rapidly every day, a number of Party and national organizations and a number of army units, in matters of radio contact, commit many serious blunders, compromising secrecy by not following exactly the instructions of the Central Party and government concerning the system of maintaining national secrecy. Such errors arise from a situation in which a number of systems have yet to be settled, but especially from cadre at all levels, all branches in charge, lacking revolutionary vigilance, not yet perceiving clearly the schemes and tricks of the American imperialists and their lackeys."[7]

The instructions reminded the echelons and branches to carry out properly a number of particulars in the maintenance of secrecy:

1. Strictly carry out instructions of the Central Party and the decisions of the government with respect to the cryptographic task system and the administration of radio stations. Pay attention to education in security consciousness and a spirit of responsibility for maintaining secrecy on the part of cadre and personnel performing the cryptographic and radio station tasks, while, at the same time, rectifying organizational and unit systems and internal regulations for protecting secrets, making them truly strict.

2. Cease the independent issuance of cryptographic systems and the use of simple systems that do not ensure secrecy.

3. Do not send messages in the clear over radio. . . .

"The Secretariat gives the Cryptographic Section of Central [the task of] research and production, allocation, direction, and control of the use of the various types of cryptography for all organizations and units. The Secretariat delegates [authority] to the appointed Party committee of the Ministry of Public Security to administer all liaison regulations of radio stations, including top secret radio stations, together with the General Directorate for Posts and Telecommunications, to research and allocate frequencies for radio stations and to monitor the use of those frequencies, to advise, and to watch over all echelons and branches in implementing the decisions concerning the administration of radio stations."

The instructions emphasized that "the protection of secrecy in communication-liaison by radio is the number one problem of importance in the protracted war between ourselves and the enemy."

Following up on the instructions from the Secretariat, on 9 June 1966 the Prime Minister issued Instruction Number 96/TTg concerning the matter of increasing the keeping of secrecy in the use of telegrams: It said in part, "Keeping the nation's secrets is a matter of national discipline: in time of war, this discipline must be more strict."

Implementing the instructions from the Secretariat and the Prime Minister's office, the Standing Committee of the Central Military Committee issued Instruction Number 48/QDTW dated 30 July 1966 concerning "The Use of Radio Stations and Cryptography in the Army." The instruction was signed by the comrade secretary of the Main Military Committee, Vo Nguyen Giap.

The Standing Committee of the Central Military Committee pointed out that "Confronted by a situation in which the whole nation is at war, notwithstanding the daily increase in urgency of the requirements of leadership and command, and the daily expansion in the sphere of liaison, our communications and cryptographic tasks basically must continue to ensure the mission. However, the use of radio stations and cryptography by the various echelons also reveals many weaknesses, and many of them serious ones. In organizing radio nets and cryptography, there are places not yet in conformity. . . . Implementation of telegraph regulations is not yet strict; namely, sending unnecessarily long messages, wordy contents, repetitious, lacking in precision. The precedence indicator used in many cases is inappropriate. From the composition and use of covernames and operational codes, the keeping of secrecy in message content, the secrecy of radio stations, the secrecy of cryptographic systems. . . still there are many shortfalls--rather many command comrades are not yet paying adequate attention to the use of radio and cryptography. . . . "

The Standing Committee of the Central Military Committee instructed the party committees in the MRs, services, branches, organizations and units in some specifics regarding the use of radio and cryptography and issued some decisions:

All command echelons will monitor and correct communications nets for conformity.

In circumstances in which it is essential to open up radio communications then you must research closely and follow precisely the principle, if you have a radio station, you must have a cryptographic [capability].

Cease the use of cryptonyms and code words produced by unit organizations to write in secret messages.

Strictly forbid the sending of messages by radio in the clear.

Constantly educate and monitor the implementation of regulations and decisions concerning the use of radio stations and cryptography; at the same time, settle strictly the violations of principles for protecting secrets involving cryptography and radio stations.

All command levels and cadre using radio and cryptography are to research the task regulations from the Central Party Secretariat and the decisions of the General Staff with respect to the use of radio and telegrams.

Military students must pay attention in training to become command cadre, to know how to use the radio and cryptography in exercising command.

Under the concrete guidance of the party committee and the commander, operations, communications, and cryptographic organizations at the various echelons must constantly oversee, research, and rectify the task of ensuring command secrecy. The Directorate of Communications-Liaison and the Cryptographic Directorate of the General Staff are responsible for guarding and tightly monitoring the implementation.[8]

In August the General Staff organized a conference to thoroughly grasp and implement these important instructions from the Central Party Secretariat, the Prime Minister, and the Central Military Committee, the composition of the conference comprising as delegates the heads of the staff, operations, communications, cryptographic, and guard [bao ve] organizations in the Ministry of National Defense, the MRs, services, branches, and the organizations of equivalent units, etc. After the conference had researched and thoroughly grasped the instructions, the General Staff disseminated concrete decisions concerning organization and use of radio stations and cryptography, and the use and administration of secret messages of various types.

Afterward, the MRs, services, branches, organizations, and units in turn organized conferences to thoroughly grasp and expand the implementation of the instructions and decisions from upper echelons by means of concrete measures.

The Cryptographic Directorate and cryptographic organizations at the various levels helped their political commissars and commanders organize and implement these instructions and decisions. At the same time, they organized for all cryptographic cadre and personnel study sessions to thoroughly grasp and carry out the instructions and decisions of the Party and to mobilize for the technical and professional task of the branch.

Taking into consideration the instructions from above, cryptographic organizations, along with operations and communications organizations, closely maintained routine collective action in organizational tasks to ensure command and constantly trained in exercises to realize the unit missions of combat preparation and of combat. Some places, such as Air Defense-Air Force, organized annual "Command Secrecy Conferences," comprising the politico-military commanders, staff, operations, cryptographic, communications, and guard for an estimation of the situation involving the implementation of regulations, principles, and decisions and to bring out new requirements and zealous measures of implementation to produce better results in the newly developing situation.

In a number of other units, commanders also frequently announced their assessments, praising good aspects – units and individuals that performed well – criticizing units and individuals that performed incorrectly, recalling to mind the decisions or making more concrete by new decisions and putting out standing operating procedures for better execution.

OUTSTANDING ACHIEVEMENT OF THE MISSION IN THE WAR OF THE PEOPLE TO DEFEAT THE AMERICAN IMPERIALISTS' "LIMITED WAR" IN THE SOUTH

In the South, right from the opening months of 1965, the Central Directorate* [Trung uong Cuc] assessed the situation and implemented a single-minded strategy, on the basis of defeating the "special warfare" of America and her lackies, continuing to hold fast and to truly develop to take the initiative in attack, preparing in every aspect, getting ready to shatter America's large-scale counteroffensive plan for the 1965–1966 dry season.

Responding warmly and well to the fifth emulation stage and taking into consideration in their emulation communiques the entire army cryptographic branch, the Cryptographic Bureau of Southern Region HQ promulgated a plan for directly subordinate units and mobilized the entire Region cryptographic in a wave that caught up the entire military, to obtain professional knowledge as central, to serve guidance in performing the task first and foremost, to pay attention to raising the quality of the task: improvement in work style to raise productivity, improvement in work routine.

Main force troops expanded rapidly; operations were continuous and mobility high. Requirements of the organizations for support to operations were urgent, time-sensitive, and tight. Region cryptographic organized many additional cryptographic nets quickly with the expansion of the forces. The year 1965 witnessed the burgeoning of cryptographic liaison nets: the rear services net, comprising thirty-four liaison points; supply stations' net of nineteen liaison points; net for receiving supplemental military personnel, with three points; R forward area net, with twenty-five points; field combat units' net of twenty-four points; and a number of nets of units directly subordinate to R.[9]

Enemy raids destroyed cryptographic bases, and there was also concern for operational units on the move. A number of command comrades were still using many cover names and slang, making it difficult to make out messages,[10] but cryptographic cadre and personnel had to really be on their toes, enthusiastically serving timely guidance. Cryptographic of the group guarding the Region base was at one moment fending off enemy attacks on the base and at the next doing a good job of ensuring the guidance to resist the raids.

*Or Central Office, South Vietnam – COSVN – to Americans. Tr./Ed.

133

In 1965 the HQ of the Eastern Area MR congratulated the sector's Military Cryptographic Section: ". . .building a branch with a tradition of in-place study, an attitude of speed and accuracy in use, training and developing its ranks of cadre and personnel to respond as required to the expansion of forces and timely service to guidance – a tradition of withstanding hardship and overcoming obstacles, in production, labor, base construction and in the life and health of the unit."

****

The period of "limited war" was also a period in which the main-force forces in MR 5 were built up and expanded all over the place.

Regional units were also built up to a large scale, from the provincial Military Affairs Section (provincial unit) down to district and village units. Regional troops were established in each province, from one to three infantry battalions, with districts having one to two companies.

With the rapid expansion of the armed forces of the MR, organization of the MR's cryptography also had to take giant steps in order to meet the requirements to serve guidance and command at the various echelons.

In 1965 the MR Cryptographic Section became the Cryptographic Bureau of MR 5. Divisional Cryptographic Sections, regimental and provincial unit cryptographic subsections, battalion cryptographic teams, and independent company cryptographic had unified guidance concerning professionalism from the MR Cryptographic Bureau on down. Besides assistance, cryptographic cadre and personnel went down from the North, from fifty to one hundred comrades a year, yet the MR still had to send [people] off for training and development at the Sector 5 unified cryptographic section's cryptographic school, besides which, as a matter of urgency, new on-the-job training had to be done to have sufficient tables of organization and compensate for losses. Also from these years of arduous and violent combat the ranks of cryptographic cadre and personnel of the MR had grown quickly.

Applying the Party's military line to obtain guidelines for both fighting and building – building in order to serve the fighting – in the process, MR cryptographic gradually built a fully worked-out cryptonet system in its area of responsibility. The working out of a liaison net system from MR to reconnaissance and sapper companies, joint communications with friendly units, skip-echelon liaison, etc., also was a process of striving, researching, and augmentation in order to have the right code system for each strategic area [vung], each region[dia phuong], each branch and main force unit. . . .

Besides transshipping a mass of cryptographic materials and technical means sent down from the North, the MR Cryptographic Section coordinated with the Sector 5 Party Committee's cryptographic to organize and produce a portion themselves, after which they had to hand over and protect the security of the materials in very difficult conditions. In order to receive cryptographic materials, the cryptographic units to the south of the MR,such as Phu Yen, Khanh Hoa, and the Highlands units, had to go every two months, even at times in which they had to cross enemy-controlled lines and then get back to MR rear base areas. Close-in units had to go weekly, and comrades were killed along the way back to the MR to receive cryptographic materials. Having picked them up and returning, cryptographic cadre and personnel still had to find glue to seal them in a can, and bury them for protection – the more valuable the cryptographic materials, the more one had to be careful.

Although caught up in violent fighting and a weighty mission, cryptographic of Sector 5 had to ensure service to leadership in defeating two large enemy counterattacks in the dry seasons of 1965–66 and 1966–67. From the battle at Nui Thanh (Quang Nam), in which an American unit was totally annihilated, to the victories at Ba Gia, Van Tuong, Chu Lai, Dong Dzuong, Phu Yen, Binh Dinh, Son Tinh, etc., becoming famous as "Sector 5," the MR 5 cryptographic organizations – more specifically the encrypting-decrypting comrades and personnel in the combat units – tightly clung to the arteries of liaison through cryptography, ensuring secrecy for the command task in all of the dangerous situations. In the enemy attack at Ba Gia (May 1965), the cryptographic comrades of the 1st Regiment went very close to the barbed-wire fence of the Go Cao strongpoint to serve operations command.

Cde Dinh, cryptographer of the 1st Regiment, 2nd Division, along with the regimental HQ, was in an operation through the jungle when they encountered the enemy, firing and using artillery to fire into the formation. Cde Dinh was badly wounded. Knowing that he could not live, the comrade crept around trying to find a way of digging into the hard earth to bury and secrete the cryptographic materials, then, making an unusual effort to crawl down a small stream gully some twenty meters from where he buried the materials, he died there.When the regiment organized a search for Cde Dinh, they discovered all of the materials which the comrade had concealed. The comrade had set an example of absolute loyalty, even though giving up his life, but not permitting secret materials to fall into enemy hands. The comrade was awarded the Order of Liberation Feat of Arms, third class.

By 1966, cryptographic liaison nets had increased very rapidly,[11] especially the net of the Military Committee and Region HQ and directly subordinate elements, which rose to fifty-four points and the internal Region net, with ninety points. The total number of points on the Region-wide network was 1,156 points with cryptographic liaison arrangements.[12]

As for technique, generally speaking, about half of the units used the "scrambled" ["lon xon" – presumably two-part code] system (principally MR 9), a fourth used handbook systems (regions 8 and 9), and a few remaining units used spell-charts.

The cryptographic organizations in the South also had to research and produce many types of dictionary systems and mixed key by themselves.

135

During 1966 the Technique Research Section of the Region Cryptographic Bureau (which was established in October 1963 with 1st Lt Manh Phuc Sanh as chief) researched and produced 136 types of dictionary codes and 1,384 sets of cipher key. (Also that year the Cryptographic Directorate assisted the South with four types of system of the KTB4 model.)

In November 1968, the Cryptographic School of Region HQ was officially approved for establishment, carrying the designator H8. Cde 1st Lt Dao Trong Loi and 1st Lt Nguyen Duc San were in charge, the mission being to prepare cryptographic personnel and give cryptographic refresher at the platoon level for the entire Region (less Sector 5 and the Highlands). In 1967, the school trained twenty-one comrades. In 1968, the school enrolled many, training 181 mobile-element comrades of the bureau, integrated into the school.

The MR 5 Cryptographic Bureau and the Highland Front Cryptographic Bureau (under Cde Do Bong as bureau chief) also implemented the mission of training new personnel. From the rainy season of 1965 to the rainy season of 1967, the cryptographic organization of the Southern Liberation Army passed through many difficulties and trials. Base areas were violently attacked by infantry, aircraft, and armor, by bomb and bullet, and by chemical substances. Along with command organizations, the cryptographic organizations had to move frequently. The Cryptographic Bureau of Region HQ moved tons of technical means and equipment by human means and did so safely. The cryptographic organization of the Region and the MRs had to be split into many detachments to serve in many CPs. Cadre and personnel routinely had to work in underground shelters deficient in air and lighting. In areas occupied by the enemy, the fellows lived and worked in secret underground shelters, or in underground passageways. In the Mekong delta, the fellows studied and worked on battlefields with many canals and irrigation ditches, immense fields of water, encountering complications and difficulties in preserving technical means. In the Highlands, almost all cadre and personnel suffered from malaria, sapping their health. In the fighting to beat the enemy back, many comrades were heroically sacrificed (in 1966 alone, twenty-nine comrades were killed). Ensuring command secrecy over many cryptographic liaison nets that were broad and deep, with complex variations, the Liberation Army's cryptographic organizations had to meet many requirements to link-up connections on an emergency basis. Message volume grew larger daily. And in these times there was much lack of cadre and personnel. The old comrades (for the most part, up in years), suffered from disease, constantly enfeebled, unable to do a good job of ensuring the task under circumstances of endless hardships – the new comrades lacked experience, needed time to become acquainted with the environment, etc.

With steel-like confidence in the ultimate victory of the revolution, in the spirit of "because the people are self-sacrificing, sacrifice to save the nation," the cryptographic warriors of the Liberation Army were united, resolved to exert themselves to strive to go forward to fulfill the task missions of training, labor, production, and combat. Many examples of dedication to the task and heroic sacrifice occurred.

Cde Ho Minh Khan, an intelligence [tinh bao] cryptographic warrior in the enemy's rear, when captured by the enemy, displayed high revolutionary courage in the face of savage torture and their subtle enticements. Not saying one word, Cde Khan was victorious and a lofty sacrifice.

Comrade Dinh Kim Sung, a cryptographic warrior in the 1st Regiment, 2nd Infantry Division of MR 5, in a battle at Tay Son Tinh (Quang Ngai), along with his team, fought off numerous enemy attacks on the regimental CP, protecting two comrades who carried cryptographic material out past the enemy encirclement to safety. Cde Sung shot and killed three Americans and with teammates brought down one American airplane. Heavily wounded, the comrade continued to fight and died heroically on the field of battle.

In April 1967, the cryptographic team of the Kien Phong Provincial Unit was raided by the enemy. The comrades separately buried the technical means and cryptographic key in one place and the code in another, and fought until the last breath.

On the night of the 30th of January and early hours of the First of Tet in the Year of the Monkey (1968), the offensive broke out simultaneously in sixty-four cities, villages, and hamlets and regions of the countryside. The army cryptographic branch fully performed its duty, contributing an important part in the work of ensuring the elements of secrecy and surprise for the offensive and uprising. It especially did its part in ensuring absolute secrecy of objectives and the preparation period throughout the theater. From upper organizations of staff and strategy to the MR, divisional, regimental, battalion, etc., organizations, the cryptographic organizations all combined ensured secrecy, accuracy, and timeliness in the contents of leadership, guidance, and command in all points of contact. The comrades assigned in the units thrusting deeply into the cities, [those] in reconnaissance units, sappers, in the Saigon front, the Tri-Thien-Hue front, and the Khe Sanh front, the cryptographic organizations of the 308th, 304th, 320th, and 341st divisions and other units, including the first tank unit to appear in the South, having overcome fierce hardships, ensured command secrecy in 170 days and nights of continuous fighting, playing their part in the great victories of our army and our people at Lang Vay, Ta Con, and Khe Sanh.

During the time of the general offensive and uprising in the South, the number of cryptographic liaison nets in the Bureau of Encryption and Decryption of the Cryptographic Directorate increased by 60 percent; outgoing and incoming messages with the theater shot from 5,000 official messages a month up to 13,000 official messages, high precedence.

The offensive and uprising of our army and our people won resounding victory, thus upsetting the strategy of the American imperialists. The limited war strategy had failed completely; the enemy had fallen into a totally defensive posture. Likewise suffering stinging defeat in the war of destruction in the North, American President Johnson had to announce a limit in bombing, then a total cessation of bombing, with no conditions on the part of the Democratic Republic of Viet Nam, and had to agree to a quadripartite meeting in Paris.

The army cryptographic branch played a worthy part in the victory of our army and our people, especially the cryptographic organizations in the South. The cryptographic cadre and personnel not only served combat command but also fought in self-defense, to protect the security of cryptographic technique, and to protect command organizations. Side by side with that, they also had to shift for themselves for messing, in ensuring survival, and in participating in tasks of other organizations. Cryptographic cadre and personnel were very honored and took pride in having done their part, blood and bone, in this greatest of victories. According to incomplete records, in the campaign forty-one comrades were valiantly sacrificed and thirty-nine wounded. Among them were many role models for the entire branch to study and copy: Cde Nguyen Van Giai, intelligence [tinh bao] cryptographic warrior in the Cu Chi "earth of steel," with a high sense of purpose protected secret Party materials to the end, keeping them from falling into enemy hands. Cde Nguyen Van Dau in Western Nam Bo, when he fell into an enemy encirclement, fought courageously and concealed the cryptographic materials before being killed. Corporal Nguyen Van Thang, cryptographer at Military Relay Station 35 of HQ of the 559th, served on the Ho Chi Minh Trail; as he was going to hand off cryptographic materials for the relay station, he tripped an enemy mine, badly wounding him, taking off both legs; his entire body wounded, knowing that he could not live, the comrade withstood intense pain to wrap up all of the codes, key, and cryptographic materials in his mosquito net, then set them afire before dying, protecting the secrecy of the materials.

Cde Van Tien Dzung, Chief of the General Staff, in congratulating the cryptographic branch on the occasion of Spring 1968, said: ". . . recently the cryptographic brothers and sisters strove to overcome every difficulty to do a good job of accomplishing the mission of ensuring secrecy and timeliness for contents of leadership and command, securing victory in the theaters. Representing the Central Military Committee and the High Command, I commend the comrades' efforts and accomplishments."

The Military Committee and Region HQ also highly valued the results of the task of service by the cryptographic organizations in the Southern theater.

Notes

1. The number of cryptographic cadre and personnel increased rapidly. In 1965, cryptographic of the Air Defense-Air Force Service increased by 167 percent. In 1966 the MR 4 Cryptographic increased by 44.94 percent, the Naval Service Cryptographic by 47.05 percent, GDRS Cryptographic by 129.41 percent, and the Engineer Branch Cryptographic by 141.11 percent compared to the previous year.

2. At the end of 1966 the printing plant had 146 people: 3 officers, 143 soldiers and workers.

3. The Navy's boat forces were regularly operating at sea, using radio as their number one system [of communication], thus there was insufficient cryptographic table of organization down to each individual boat.

4. The Air Force alone had seven types of opcode: the supplemental opcode, the tactical opcode, the training opcode, the pilots' opcode, the preflight opcode, the joint opcode, and the operations [hanh quan] opcode.

5. For the branches of radar (naval), communications, operations, etc.

6. Instructions No. 129-CT/TW, signed by Cde To Huu, Central Party Secretary, preserved in the Cryptographic Bureau.

7. Extract from Directive No. 129-CT/TW.

8. Extract from Instructions No. 48 of the Central Military Committee.

9. The 16th Regiment (Sector 5 entering) in contact with R, Sector 5, Western Area Front HQ, artillery regiment, guard regiment.

10. For example, R Forward ordered the 3rd Regiment: "E5 [5th Regiment] across the Be River is caught up in flooding. E3, help organize to get E5 over the river." When E3 crypto decrypted, because there were garbled secret designators, they could not understand the message content to correct it.

11. MR 7 increased to 33; Saigon-Gia Dinh Special Sector increased to 8 points.   MR 8 increased to 52; Region 6 increased to 15.   MR 9 increased to 30; Region 10 increased to more than 200.

12. In the total of 1,756 points for the entire cryptographic branch in Theater B.

Chapter Seven

Ensuring Service to Leadership, Guidance, and Command in Defeating the American Imperialists' "Vietnamization Strategy" and their Second War of Destruction in the North (1969–1972)

After the 1968 Tet general offensive and uprising of our military and people in the South, the "limited war" strategy of the American imperialists had failed completely. In order to continue to preserve America's neocolonialist system in the South – in order to cope with the whippings and powerful attacks by our army and our people – Nixon promulgated the strategy of "Vietnamizing the war." Aiming at carrying out this stratagem, the American imperialists accelerated "pacification" of the countryside; pushed the build-up of the Saigon puppet army to create a modern army to gradually replace the American military; and at the same time used coordinated methods with respect to military, economic, foreign policy, political matters, etc., drastically counterattacking and broadening the war throughout Indochina.

Facing the war schemes and operational tricks of America and her lackies, the struggle by our army and our people to repel the armies of aggression intensified, to the point that it was inexhaustibly tough and decisive.

In April 1969, the Central Party Executive Committee issued a resolution to mobilize the power of our entire military, our entire people, to expand the strategy of attack, to beat the "Vietnamization" scheme, to beat it so that "Americans out, puppets collapse" proceeded to secure a decisive victory.

In this situation, the PAVN cryptographic branch stood facing requirements and extremely weighty, complicated missions. Implementing the instructions of the Central Military Committee, the army cryptographic organization drew up timely plans and procedures to carry out the specialty tasks responsive to the requirements to serve leadership, guidance, and command in the new situation.

The army cryptographic organizations in the North concentrated their efforts to implement these missions:

- Organize to serve leadership, guidance, and command secrecy, accuracy, and timeliness in every set of circumstances, simultaneously continuing to accelerate heavily the work of changing over to employ techniques KTB5 and KTC. The mission was fixed as the number one central mission.

- Step up the task of enrolling students for development; give refreshers to the cadre and personnel.

- Zealously organize people to assist and provide technique for cryptographic in the South, Laos, and Cambodia.

- Increase the task of professional guidance, thoroughly grasping the situation, dealing with timeliness and accuracy in the developing situation, thereby responding to each mission requirement.

Vis-a-vis the cryptographic organizations of the Southern Liberation Army, the Southern Military Committee [Quan uy mien Nam] instructed:

"Mobilize to the highest level the power of all cryptographic cadre and personnel; strive to bring into play the achievements that have been accomplished; resolve to overcome unresolved errors, valiantly, and with dedication to the task and to fighting; build the organization, pure and solid; never cease to raise the level of cryptographic technique, the level of use of technique and of professional ability; strictly implement the cryptographic branch's table of organization and assignments; aim at serving the leadership, guidance, and command of the Military Committee, Region HQ [Bo chi huy Mien], of political commissars and commanders at all levels, through secrecy, accuracy, and timeliness."

Implementing the above instructions, the Cryptographic Bureau of the Region's Military HQ [Bo Chi huy quan su Mien] organized to carry out five major task aspects:

- To increase the task of political education, ideological leadership, and administration of cadre and personnel, to create in the ranks of cadre and personnel a thorough grasp of the strategic determination of the Party – a steadfast class outlook, overcoming the mind set of a fear of hardship – laying down one's life, bringing into play the feelings of valiantry, the spirit of dedication to the task, in labor, study, and combat.

- Step up the task of technical and professional training, with the line, "make on-the-job refresher the center; get training in the real world task as the essential," raising the capacity to execute the mission in every situation.

- Make an effort to overcome difficulties, seriously implementing the policy of changing over to the use of technique KTB 5. Increase the administration of technique, for security and secrecy.

- Build into routine a thorough grasp of professional guidance from top to bottom.

- Organize the task of encrypting and decrypting messages to serve leadership, command, and guidance, so that it is good in every circumstance.

From 1969 on, our army and our people stepped up the counteroffensive and attacked the enemy. Requirements for the army cryptographic branch to serve leadership, guidance, and command demanded a larger dimension. There were many days-long campaigns and combined-branch operations at a high level, among which were, as examples, campaigns such as the Route 9-Southern Laos campaign, the general strategic

142

counteroffensive campaign of 1972, the campaign to defeat the strategic attack by America B-52s on Hanoi and Haiphong in December 1972, etc.

Also, from the end of 1969, our army had expanded greatly with respect to organization, the 559th Group adding the 470th Division to cooperate with the Highlands Front so as to greatly expand the transportation route down to Eastern Area Nam Bo. By mid-1970 there had been integrated, in addition, the 968th Front and the 565th Specialty Group (with the HQs of sectors 470, 491, 472, 473, and 571). By the end of 1969, the sapper troops in the South had also expanded all over the theaters. By 1970, the branches – artillery, armor, sappers, communications, engineers – had all stepped up a notch in expansion. In October 1970, Groupment [binh doan – quasi-corps] 70 was formed, comprising the 304th, 308th, and 320th Divisions and regiments, battalions, and branches, in order to meet the operational requirements for combined branches in large campaigns.

Faced with the operational requirements and the expansion of armed forces in the new period, the system of organization and alignment of the ranks of cryptographic cadre and personnel, and the cryptographic-technique liaison net system, took a big step up.

From the army cryptographic organizational standpoint, as of 1972 there were 4,755 units with cryptographic organizations in the North; in MR 5 there were 600 units with cryptographic organizations; and 1,962 in Nam Bo. Cryptographic organization in the Highlands and MR Tri-Thien had also expanded greatly.

Along with the expansion of the system of cryptographic organization, the system of cryptographic technique also expanded steadily in depth and breadth in the command organization system of the armed forces.

In the Cryptographic Bureau of Region HQ in 1969, the number of liaison points which the bureau had to cover was 76, rising to 128 in 1970, and 154 in 1972.

By September 1970, the cryptographic liaison system between Central and the Central Military Committee had organized liaison directly to MRs 6, 8, and 9.

In the Encrypting and Decrypting Bureau of the Cryptographic Directorate of the General Staff, the number of points for which liaison had to be ensured in all three theaters was 157 points in 1969; 214 points by 1972; and 341 points in 1973 (that had liaison).

The system of cryptographic organization and the system of cryptographic liaison nets expanded greatly, requiring a corresponding quantity of cadre and personnel to meet the requirements for ensuring communication for the theaters. The army cryptographic branch had to strive to exert itself to the utmost to build and to develop the ranks of cadre and personnel to serve in the tasks of leadership, guidance and command of the Party and the army.

The average yearly number of army cryptographic cadre and personnel increased from 40 to 50 percent. MR4 increased by 280 percent; the Air Defense-Air Force Service by

258.97 percent; HQ of the 559th by 190 percent; HQ of the 959th by 227.58 percent; and the 316th Division by 537 percent. Figured to November 1972, the grand total of army cryptographic cadre and personnel had risen to 5,337 comrades.

The development of cryptographic cadre and personnel during these years became a large mission and an urgent one for the army cryptographic branch. Because they had to be arranged in spots where security and secrecy could be assured, classes for developing cryptographic personnel in the North encountered many difficulties in teaching, study, and organization to ensure living conditions. Side by side with education in political matters and ideology, and in the speciality professional techniques, there was special importance on training in physical conditioning, to get into shape for the march to distant theaters and to build endurance for the conditions in which the encryption-decryption task would be performed in places of hardship and violence later on. The curriculum content for teaching the specialty techniques was continually upgraded. The program of study reserved appropriate time for the student to practice technique and apply it in the field.

There were emergency classes in which the program and time had to be curtailed, but many students continued to achieve high scores in the subjects.

The development of army cryptographic personnel in the Southern theater also encountered difficulties from many aspects, from the selection of students to the task of ensuring living conditions for the school in situations of violent fighting. Most warriors going down from the North had to wait until the General Staff Cryptographic Directorate investigated and made a determination as to the implementation of developmental organization. In MR 9, they established a source of enrolling students by seeking younger siblings of cadre, from the masses, hard-core revolutionaries, to perform liaison or enter a technical materials transportation unit of the MR cryptographic; through the process of nurturing, educating, and training, they were put to the test, admitted to the [Party] group, the Party, then sent off for cryptographic training. With this method, the MR cryptographic organization had rather taken the initiative vis-a-vis sources of enrollment, meeting a third of the troop strength for development and a quality that was also guaranteed to be better.

The Southern Liberation Army Cryptographic School, although small in size and not yet having a regular routine for development, nevertheless exerted itself fully to continuous development, not reckoning "courses" and "classes," opening sessions at times with only fifteen to twenty or even ten people, many times having to hold two one-hour classes consecutively.

In MR 8 and MR 9, there were times in which the cryptographic classes had to be moved two or three times. Students were entirely on their own, having to be self-supporting, with respect to living conditions, paper and ink, etc. The comrades sought to overcome obstacles by means of having one class session under way while another went fishing to get fish to sell to buy rice to eat and things for study. There was a time when a heavy enemy strike devastated the class and the students had to sit in the shade of a tent-fly under a paperbark tree to study. Cde Tam Ky slipped out alone to buy rice for the class, was ambushed by the enemy, fought, and died heroically.

From 1969 until 1972, the Army Cryptographic School graduated 2,389 students. From 1969 to 1971, the Southern Liberation Army Cryptographic School graduated over 250 people. MR 9's class graduated sixty-five people.

Counting 1965 through 1972, the army cryptographic branch graduated over 8,000 personnel, including more than 600 women.

The Military Cryptographic School in the North graduated more the 6,000 personnel; the cryptographic school classes in Nam Bo graduated 928 personnel.

The Sector 5 and Highlands training classes graduated nearly 1,000 personnel, etc.

Thanks to these numbers trained from 1965 to 1972, 3,583 army cryptographic cadre and personnel supported the battlefields of the South.

FINISHING THE CHANGEOVER TO TECHNIQUE KTB5 AND EXPANDING THE USE OF TECHNIQUE KTC

Concurrent with the task of training the ranks of cadre and personnel, the army cryptographic branch continued to press strongly to effect the changeover to the use of technique KTB5 and started to develop the use of technique KTC at the principal points.

After the December 1968 conference on training and the use of technique KTB5, the Cryptographic Directorate of the General Staff directed the army's cryptographic organizations and activities in the North and the Volunteer Army in Laos to expand the use of technique KTB5.

Early in 1969, after just two months, the cryptographic organization of MR Tri Thien had totally changed over to technique KTB5, right down to the lowest units, although the responsibility for serving command and combat was very tense.

In the Laotian theater, with the exception of some units behind enemy lines and caught up in combat, and which had not received the new type of technique, the Vietnamese Volunteer Army's cryptographic organization also essentially accomplished the changeover to the use of technique KTB5.

On 15 December 1969, the nationwide, army-wide cryptographic cadre conference summarized the situation of changeover to use of the new technique and clearly stated:

"After just a little more than a year (September 1968- November 1969) we essentially accomplished the changeover to the use of technique KTB5, replacing KTB4, with a raised standard of use,"[1] at the same time continuing to ensure every respect of leadership, control, and command in the fierce fighting of the new period,etc. "This achievement was very great, and the greatest of all was raising the level of cryptographic technique a step, with profound implications in the struggle to counter the American imperialists' gleaning

145

of information through cryptanalysis, opening up a pleasant prospect for us in realizing our course of developing technique."

As to the reason for success in the changeover to the use of technique KTB5, the conference analyzed this, and concluded: First of all, it was because of direct leadership from the Central Military [Party] Committee, and from the concerned chiefs – guidance and help in every aspect: The cryptographic branch had the direction to develop, and the policy of changing to, the use of the new technique precisely, with guidelines and suitable methods, applying the experiences of the time of changeover from technique KTA to technique KTB4. The Cryptographic Section of Central and the cryptographic directorates of the Army, Central Party, and Public Security had concrete plans of action, and creative methods of execution. Cadre of the branch were of one mind, striving to surmount difficult obstacles in research and production, entrusting responsibilities for training and the use of the technique to exemplary cadre. Many comrade bureau chiefs and section chiefs made the effort themselves to gain mastery of the new technique in order to directly train subordinate cadre and personnel.

Also at this conference the delegates brought up some mistakes in the changeover to the new technique. From the standpoint of ideology, we still had comrades who revealed a subjective attitude for technique KTB4 as offering the very highest level of security, negating the need to change over to the use of technique KTB5. There were comrades who doubted that the KTB5 technique was all that reliable, fast, and accurate. There were units that had only developed the use of technique KTB4 to accomplish their mission. Attitudes that recoiled from difficulty or mental stress were revealed. Concerning guidance, there was also subjective thinking, not yet anticipating fully the difficulties, thus still lacking in thoughtfulness and closeness in ideological leadership, in building with determination in training and use, still not paying strict attention to the degree to which units were weak or were still having excessive difficulties. As for research and quality production of the various types of dictionary codes and random key, those too were not high. Through theoretical analysis and results in actual use of KTB5, the conference concluded: "KTB5 not only has a higher level of security than KTB4, but has better error correction and accuracy assurance than KTB4. Thus KTB5 serves well in meeting each requirement for leadership, guidance, and command in every situation of an unexpected nature, in missions at any echelon, in any branch, any theater, with respect to strategy, campaigns, or combat."

In 1970, the Central Military Committee directed the MRs, arms and services, the various organizations and units, to pay attention to providing close guidance and creating every condition for the army cryptographic branch to implement training and use of the new technique favorably.

The changeover to KTB5 in the Southern theater encountered many difficulties because of combat service that caused large volumes of work and of urgency. Because support in the form of types of system and cryptographic key was inadequate, many units received permission to produce their own technical means, such as Cryptographic of Southern HQ [Bo tu lenh mien Nam], MR 5, MR 9, etc., but production also ran into snags, principally from the standpoint of funds. The distribution of cryptographic materials in a split-up war theater situation also made coming and going difficult and dangerous. In MR 9, it could take one to two months [to get] from the unit back to the MR. With resolve to overcome difficulties and hardships, units in the South urgently implemented a program to use the new technique. By the beginning of 1971, army cryptographic units in the South had accomplished the changeover to the use of technique KTB5 in all units.

In March 1969, the Central Cryptographic Committee convened a discussion of methods of expanding cryptographic technique, looking mainly to changing over to the use of the KTC technique.

In essence, technique KTC is a type of cryptographic technique with a level of security and high degree of accuracy and speed, valuable as a good type of cryptographic technique of the randomized type of cryptography, but also with many complexities. It demands a tight organization plus time and effort, and must have a refined level of use. This is a major difficulty with implications for the entire process of steps – research, production, training, use.

People who did encrypting and decrypting by technique KTC5 had to give up on many strong points, had to stretch their brains more, had to stand their ground, had to endure more, compared with the other previous types of cryptographic systems.

In July 1969, the Cryptographic Directorate of the General Staff organized cryptographic cadre training to implement a plan for the training and use of technique KTC.

Cadre trained in the use of technique KTC returned to their units as core cadre for training the units in the new technique. Units in the South had their cadre and personnel interchange to consolidate training in the new technique under conditions of alternating study and carrying out combat duties.

From 15 September 1969 the cryptographic organizations at the level of MR and service in the North, MR Tri-Thien, MR 5, and HQ, Southern Area implemented experimental use of KTC3 and KTC5 with the Encryption-Decryption Bureau of the Cryptographic Directorate of the General Staff.

The army cryptographic branch mobilized an emulation campaign, studying and training in the new technique in school and in the office, in the spirit of "Resolve to Defeat the American Aggressors."

Warmly responding to this campaign, cryptographic cadre and personnel army-wide, from rear areas to front lines, from mainland to distant sea islands, and even into the heart of the enemy, in concentrated spots and scattered and independent task teams, raised an atmosphere of eager bustle, training industriously in the new technique, showing creativity in study and training, many units and individuals achieving high productivity and quality of encrypting and decrypting, some comrades setting records.[2]

Army cryptographic cadre and personnel had to be continuously unruffled, had to know that they were in control in the task, and had to almost continually come into contact with extremely secret and important problems of the Party and the army, through encrypting and decrypting message contents, including no few messages containing news of victories that warmed the cockles of their hearts.

But none of the cryptographic cadre or personnel could repress their feelings, when suddenly, in the days toward the end of August 1969, they had to encrypt and decrypt the message from the Central Military Committee transmitting a communique from the Central Party Secretariat concerning the health of revered and beloved Uncle Ho. The entire text of the message sent at 2230 hours on 29 August is as follows:

From the Central Military Committee to the comrade secretaries of the Military Region committees, the division committees, independent regiment committees, and equivalent units: The Military Committee is passing to all comrades the communique of the Secretariat concerning the health of Uncle Ho. All comrades will at once organize announcements in the committees at the various echelons in strict accordance with regulations and with due regard to ensuring absolute secrecy.

VAN [Vo Nguyen Giap]

The message of the Secretariat (Top Secret message) [dien tuyet mat]:"Per resolution of the Politburo, the Secretariat begs to communicate concerning the situation of the health of President Ho as follows:

The entire Party – all of our people – know that our Uncle Ho was originally in very good health; Uncle regularly kept an eye on physical training and worked on an organized, planned basis. As a result, he was able to come through many hardships, imprisonment and exile, disease--even though up in years, Uncle continued to have the strength to shoulder every responsibility the Party and the people placed upon him. But over the past decade, Uncle's health had begun to decline. From 1965, there were many times in which Uncle suffered from dangerous attacks. Often we at Central organized the work with solicitude and consideration, especially relying on Uncle's strong efforts, but these attacks persisted. 'From the beginning of 1968, the Politburo reorganized methods of working, so that Uncle could take part in the discussion of the major undertakings of Party and State, sometimes getting involved in important activities, while at other times conserving Uncle's health. But from the beginning of the year until now, Uncle's health continued to decline. This August Uncle suffered a drawn-out attack (some of the bouts were critical), and has not been well right up until today.

At present, the Politburo is concerned with organizing Uncle's cure, and is confident that, as was the case each time, Uncle will prevail. But because the problem of Uncle's health is a problem of great importance to all of the people, and to all of the Party, the Politburo made a thorough report on the above situation to the Central Party Executive Committee and this status report is communicated within Party circles to the standing committees of the Sector committees, the city and provincial committees, the Party committees [Ban] and groups, the standing committees at the Party Committees [uy] directly subordinate to Central; and, in the army, to the standing committees of the division committees and echelons comparable to division, and to standing committees of regimental and independent regiment committees. 'It is desired that comrades organize the communication [of Uncle's situation] exactly as determined by the Politburo, and keep confidence in their hearts over Central's nursing and treatment of Uncle. We love Uncle tenderly, so we must turn that sentiment into an all-out effort in fighting, in production, and in our tasks, an all-out effort for each person in their mission, each unit, each organization, as Uncle continues to expect of us as a matter of routine. Central thinks that, if new achievements in battle, productivity, and task assignments of all places be reported to Uncle at a time when he is weak and tired, then surely Uncle will be gratified.

Finally, it is desired that comrades keep this news absolutely secret, seeing it as a matter falling within the province of highest secrecy critical to the Party and the Nation.

On behalf of the Secretariat,
Le Van Luong

These messages brought tears to the eyes of cryptographic cadre and personnel over succeeding days: Uncle Ho had bid us farewell! A grievous loss beyond measure for the whole Party, for all of our people, for our entire army!

Cadre and personnel of the army cryptographic branch seriously implemented the appeal of the Politburo and the Central Military Committee to "Turn grief into strength," becoming day by day more industrious and innovative, quietly relaying messages of the Party, the army, the entire military in the months and the stages of "actions to repay Uncle," achieving worthy accomplishments consistent with the teachings and the heart of Uncle and the army and the unit.

ENSURING SERVICE TO LEADERSHIP AND COMMAND IN ATTACKING AND COUNTERATTACKING THE ENEMY'S "PACIFICATION" SCHEME

Moving into 1969, our army and our people continued to attack and counterattack the enemy on the fronts in the Southern theater, an example being the two-stage spring and summer actions.

The cryptographic organization of the region's military HQ [Bo chi huy] and the cryptographic organizations of the 5th, 7th, and 9th division, MR 7, etc., served command in the attacks on the enemy on the axes Tay Ninh, Binh Long, Bien Hoa-Long Khanh, etc., and served command in repulsing the enemy's mopping-up operation in the Dzau Tieng sector.

The cryptographic organization of MR5 and the cryptographic teams of the 2nd and 3rd divisions served command in attacking the enemy in many places, such as An Hoa, Tien Phuoc, Tu Nghia, etc.

The cryptographic organization of the Highland Front served command in the campaign attacking the enemy at Doc To.

The cryptographic organization of MR Tri-Thien and the cryptographic organization of the 324th Division served command in beating back the enemy's attack, securely protecting our strategic lines of transportation in that sector.

The cryptographic organizations of elite [tinh nhue] troops served command in striking the enemy in many glorious actions, causing the enemy much heavy loss, as in the raid on the Dong Zu base.

The cryptographic organizations of the liberation army, while serving command in counterattacking the enemy, protected the bases of Region HQ [Bo Tu lenh Mien], MR 5, MR 8, MR 9, etc.

In February 1970, the allied Laotian-Vietnamese army opened a campaign to crush the "Cu Kiet"* occupational operation of the Americans and their Laotian puppets in the Plaine des Jarres sector, the cryptographic warriors of the Vietnamese Voluntary Army right down as far as bases [co so], objectives, directions, commanding positions – all in the region of the enemy and the Plaine des Jarres command. The cryptographic organizations of the air defense, armor, artillery, sapper, engineer, air force, etc., units from Viet Nam crossed over to fight on a combined-branch basis, coordinated closely with the cryptographic organizations of the Laotian Volunteer Army to ensure that command was grasped thoroughly and completely. Cde Nguyen Van Binh, cryptographer of the Vietnamese Volunteer Army, bravely lost his life in this campaign.

In March 1970, the American imperialists organized and staged a coup d'etat in Cambodia, overthrowing Prince Sihanouk and bringing Lon Nol to power, at the same time bringing American troops and Saigon puppets into an attack of aggression on Cambodia, aimed at wiping out each and every revolutionary organization in the South, wiping out the Region's main force units, destroying our rear bases and cutting our strategic assistance transport lines, as they reverted to the "Vietnamization" strategy. Army cryptographic organizations sent along instructions from the Central Party concerning service to the leadership and command of the Central Office and MR 5 stepped up attacking the enemy and widening the liberated region along the border, helping the revolutionary movement of our friends.

In April 1970, the Cryptographic Bureau of Southern Region Military HQ (Bo chi huy quan su Mien), the 1st, 5th, 7th, and 9th divisions, and the Highland Front cryptographic organization served command in striking the enemy in eastern and northeastern Cambodia, the Highlands, and Lower Laos, crossing over and participating in glorious feats of arms that liberated six of our friends' provinces.

---

*Vietnamese rendering of Laotian "Ku Kiet," meaning "to regain, or restore, prestige." *The 316th Division*. Vol. II. Hanoi: PAVN Press, 1986, 121, 121n, and 132, although that history implies that the operation began in mid-1969. - Tr./Ed.]

In January 1971, the Americans and their puppets opened operation "Lam Son 719," striking the Route 9-Southern Laos sector with a large force: More than 30,000 main force puppet troops, with air forces and 10,000 American troops in support, aimed at severing our strategic line of transportation. Cryptographic organizations of the Front HQ and those of the 308th, 304th, 320th, 324th, and 2nd (MR 5) divisions, Group 559, HQ B3, the cryptographic teams of the battalions of sappers and of tanks, the regiments of artillery and of engineers, etc.,ensured secrecy, timely deployment of forces, and a coordinated strike on the enemy. This was a campaign in which joint operations of branches was on a large scale and over many days, requiring the organization of guidance, use, and the organization of cryptographic nets with many complications, with many wide direct-contact nets, and skip-echelon nets, but the cryptographic organizations carried out the mission in good order.

In the Cambodian theater, after serving command in defeating the aggressive attack by the Americans and their Saigon puppets, the cryptographic organizations of the 1st, 7th, 9th and 5th divisions went on to serve command in striking the enemy, inflicting heavy losses on them, an example being the counterattack that hit the enemy opening his "Total Victory" operation in February 1971.

At the beginning of 1971, cryptographic of the Highland Front had served command in counter attacking the enemy when he opened his "Quang Trung 4" operation, striking the Kontum sector with the aim of destroying bases and severing our strategic lines of transportation, while cooperating with the "Lam Son 719" operation in the Route 9-Southern Laos sector.

In the period from 1969 to 1971, the cryptographic organizations of the Southern Liberation Army performed their mission in conditions of extreme difficulty and hardship. The enemy increased his violent attacks on our rear bases and CPs. Nearly all units in the region adjacent to the enemy (in Sector 8) were struck many times by enemy planes and artillery, and had to encrypt and decrypt in underground shelters and had to frequently move – units such as Kien Tuong, An Giang, and My Tho, and the 1st and 2nd regiments. There were times at the end of 1971 in which the MR 9 cryptographic organization was only a few hundred meters from the enemy, bombs exploding and artillery firing continuously, day and night. Besides completely accomplishing the specialty mission, there were units that had to devote 50 percent of their time to foraging. A number of places had to subsist on gruel, or eat jungle tubers instead of rice. The number of cryptographers captured, missing, or casualties during this time was rather large. During the two years, 1969 and 1970, in Nam Bo alone, seventy-eight comrades gave their lives and eighteen comrades were wounded.

Cryptographic cadre and personnel of the Southern Liberation Army withstood every hardship, sacrificing themselves, resolved to do a good job of accomplishing the mission of service to leadership, guidance, and command.

So as to have enough technical materials, the Region military cryptographic organization – besides transshipping and distributing nearly 300 types of codes and 1,300

Crypto receiving its mission at Group 365, Air Defense Service

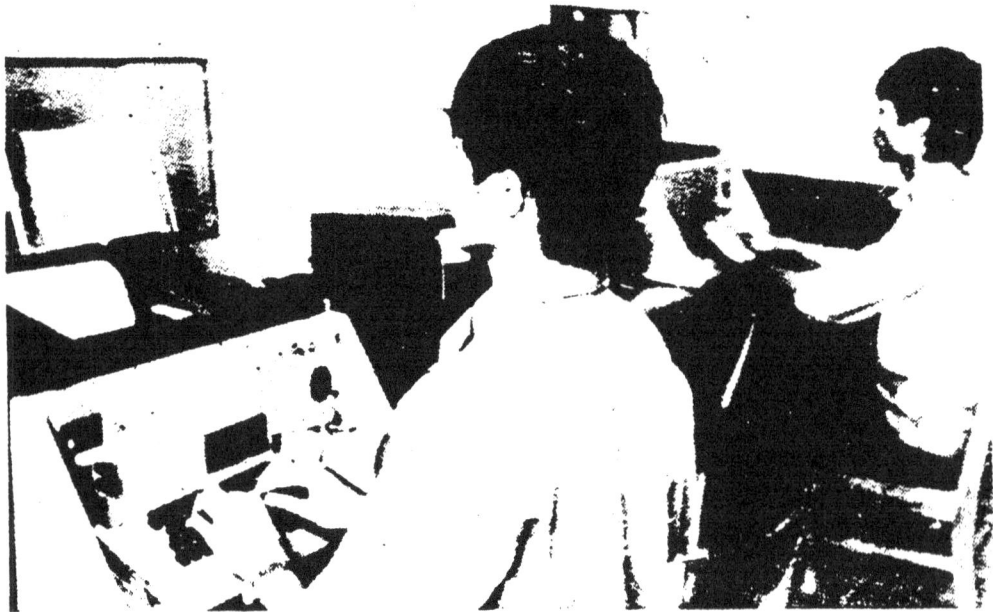

Cipher machine use in the Air Force Service Crypto Bureau

Navy crypto prepares to embark for the islands

153

sets of cryptographic key-- zealously produced on their own, cryptographic key to supply to the fronts. "With the requirement for cryptographic material rather large, and assistance limited, the bureau technique cadre made an all-out effort and did a good job of implementing the branch resolution to produce and ensure sufficient material for an expanding net, with service requirements high. . . ." "In difficult circumstances, lacking gear, lacking money to ship and buy means or raw materials, the enemy pounding the bases, they had to move supplies tens of times, undergoing bombs a half dozen times, but the cryptographic cadre and personnel press-printed thousands of sets of key of better quality, organized to transship nearly 100 stages of cryptographic materials for nearly all units in the entire [Southern] Region, administered tons of cryptographic materials securely, their distribution registered, arranged precisely, strictly per standing operating procedures." "Some units had only a few cranky machines, with three to five personnel with poor attitudes, but who also backpacked a portion of cryptographic materials for their own units, reducing assistance from above, such as at T2."[3]

The volume of encrypted messages that the Region military cryptographic had to handle during this time period went up dramatically. The number of messages in the first six months of 1969 approximated that of the entire year of 1968.[4]

In 1970, according to incomplete statistics, the grand total of messages handled by the Southern military cryptographic set-up was 735,442 official messages. During this period, [the number of] encrypted messages sent by radio and received relatively error-free was remarkable.

With the emulation slogans "When work arrives, get an experienced hand to work immediately and get it out" [and] "Until the work is finished, the heart cannot be at rest," the comrades performing the encryption-decryption task on messages worked day and night with the highest attitude, to ensure timely transmittal of the contents of leadership and command for the units.

The was also the time in which the Region cryptographic organizations were switching over to use of the new techniques (KTB5, KTC), notwithstanding the difficulties and obstacles. The average error rate of Region military cryptographic in 1970 was 1,951 in 5,564; then, in 1970, they went to 1 in 9,501. Cde Bui Thanh Xuan of the Region Cryptographic Bureau took care of sixty official messages in one day; in the first six months of the year, he worked out 3,200 official messages with 100 percent accuracy.

While performing the mission of service to unit leadership and command, the cryptographic comrades set many examples of courage in combat and self-sacrifice of life to protect the secrets of the Party and the army. In February 1969, Cde Ngo Van Hop, cryptographer of the 559th Group – although during a time of enemy B52 bombing – fearlessly crossed through fire and shell to deliver a secret message to the command in timely fashion, and he courageously gave his life returning to his unit. In MR 9, during a phase of serving combat command in striking the enemy in the battle of Cha La, the underground shelter of Cde Ngoc was collapsed, and when his buddy extracted himself, the only thing – and the last thing – he said was to recommend that his buddy carry the cryptographic materials back and hand them over to the MR Cryptographic Section. Cde Nguyen Van Sy, badly wounded and near death, still used his mosquito net to wrap his materials and set them on fire to avoid their falling into enemy hands, and he died in the flames that destroyed the cryptographic materials. In Subsector 23, when the enemy was mopping up the base of the Kien Tuong Provincial Unit, the secret underground cryptographic shelter was discovered by the enemy. Three comrades, Dziep The Tai, Nguyen Van Son, and Nguyen Van Chuong, suddenly threw up the cover of the shelter, opened fire on the enemy, killing six and breaking the enemy encirclement. Cdes Tai and Son were sacrificed, Cde Chuong was wounded, but the security of their materials was protected.

In one stage of transporting cryptographic materials to be handed over to the basic units [don vi co so], three comrades in the cryptographic material transportation unit (one comrade responsible for the technical materials and two comrade guards) of MR 9 encountered an enemy ambush – the comrades struck back at the enemy with determination and protected the security of the cryptographic materials, two comrades losing their lives in the course of fighting.

ENSURING LEADERSHIP AND COMMAND IN THE 1972 STRATEGIC OFFENSIVE

The sweeping victories of the three-nation Indochinese revolution in 1971 shattered the "Vietnamized War" strategy of the American imperialists. After heavy successive losses, the American military had to fall back onto the defensive over the entire Southern theater. They hoped, with strong military blows, combined with diplomacy, to be able to force us to accept conditions that would bring them victory at the conference table and create a strong position for Nixon in the end-of-year 1972 election. Judging that we could make a large attack, America and her puppets strengthened their defensive lines and expanded their probing operations, using their strategic and tactical air forces in continuous strikes against our transportation lines, aiming to block our preparations, while at the same time pushing heavily their pacification program, striking bases inside enemy-occupied regions.

Our army and our people entered the 1972 campaign with an air of excitement from "the new situation, the great opportunity that has appeared," as the Central Party Executive Committee assessed it.

Resolved to defeat the grand schemes of the enemy, our army and our people launched a strategic offensive throughout the South, aimed at wiping out enemy strength and destroying his strong defensive lines, changing the theater situation into victory for our side, stepping up the resistance against America, and saving the nation to the final victory.

From July 1971, according to instructions from the Central Military Committee and the Chief of the General Staff, based upon a thorough grasp of the strategic objective and the intent of the campaign-- the operational blueprint, the plan for expanding forces--the

Cryptographic Directorate of the General Staff researched and prepared a plan for the campaign's cryptographic tasks, anticipated the quantities and qualities, arranged the preparation of cadre and personnel to build the cryptographic organizations, and built a network of cryptographic technique among the units participating in the operation. While focusing priorities on the preparations for launching the campaign, the Cryptographic Directorate continued to do a good job of performing the task of protecting the various theaters--the Highlands, MR 5, the Plaine des Jarres, Eastern Nam Bo, etc.

From August until December 1971, preparation proceeded, as a matter of urgency: by the beginning of 1972 the cryptographic organizations of the units participating in the campaign had basically completed the task of preparation.

In order to remedy the situation of lack of cadre and personnel, the Army Cryptographic School brought together a large number of students undergoing the prescribed curriculum and organized "lightning" classes to deal with basic problems with respect to technique and demonstration in order to augment the units in a timely way. At the same time, the Cryptographic Directorate also proposed that the General Staff second many cryptographic cadre and personnel from other units in the rear to increase the number in units participating in the campaign.

Instructions and directions having to do with the cryptographic profession in service to the campaign were promulgated, grasping the tiniest detail, thorough and complete. The task of guiding ideological and political education for the cadre and personnel participating in service to the campaign was also thoughtfully executed.

Around March 1972, the forces taking part in the campaign were posted at their groupment positions. The army cryptographic units ensured that the command task of preparation for the campaign was absolutely secret, so that the enemy would be totally surprised with regard to timing, main direction, and scale of our offensive. The cryptographic organizations of the 304th and 324th divisions and the branch regiments – although both ensuring contact and on the move in stages of long distances over months, under continuous and violent attack by enemy aircraft – still conveyed the command of the troops in timely and accurate fashion, so that they arrived at their groupment areas exactly at the determined time.

NGUYEN HUE: THE "EASTER OFFENSIVE"

At the end of March 1972, on the [Quang] Tri-[Thua] Thien front, our army carried out the general strategic offensive of 1972. This was a large-scale, combined-branch-operations campaign never before experienced in the struggle by our army and our people. Forces participating in the campaign approximated six main force infantry divisions and many regiments, battalions, and companies, branches, sappers, armor, artillery, engineers, etc., together with regional armed forces.

The task of organizing to ensure service to leadership and command by the cryptographic organizations in the Tri-Thien Campaign was also an all-out effort.

The forces of cryptographic cadre and personnel directly serving the campaign came up to 413 comrades, not counting the some 100 comrade cadre and personnel subordinate to the B4 Front. The Campaign Cryptographic Bureau, under Cde Nguyen Ngo, comprised ninety cryptographic cadre and personnel, arranged at the CPs of the campaign command. The cryptographic organization in the divisions was from twenty-two to thirty-two comrades. Each tank regiment had from eight to nine cryptographers, and the branch regiments had from four to seven cryptographers. The sapper battalions had two to three cryptographers.

The cryptonet system for the Tri-Thien campaign was extensive and complex. In the front alone, four CPs and a rear base were organized. At the primary CP, the CP in the direction of Route 1, the cryptographic organization had to ensure contact with all of the units participating in the campaign. At the CP in the northerly direction, the cryptographic organization had to ensure direct contact with units fighting in the coastal lowlands sector. At the CP to the east, the cryptographic organization had to ensure guidance for the sapper units, and the independent battalions, fighting in coordination with the entire front. In each division there were two to three CPs, while at the same time each unit had a rear base back up North. Side by side with the campaign cryptographic organizations, there were also cryptographic elements in the branch organizations – artillery, engineers, sappers, armor, air defense-air force, and military intelligence cryptographic – so the matter of linking up and sending and receiving messages was quite complicated. Campaign cryptographic organizations still had to ensure combined liaison with all of the theaters – B1, B2, B3, B4 – the MRs and the services and branches in theater A and theater C.

After proceeding with expanding the preparation, augmentation, and rectification of the cadre and personnel forces, and organizing and arranging a campaign-wide cryptonet, the General Staff Cryptographic Directorate convened a conference of cadre-in-charge at main force divisions back at the B5 CP in order to conduct a preliminary inspection of the entire task of preparation in every aspect, after which the cadre would directly inspect every unit participating in the campaign one last time. Through inspection, the Directorate mobilized the aid of the units in resolving remaining difficulties before entering the campaign, while, at the same time, confronting unit commanders with the importance of creating conditions to enable cryptographic to perform its mission completely. The Cryptographic Directorate also issued instructions and direction for the units to strive to organize study of the practical content for cadre and personnel, aiming at raising the quality of ensuring the mission ahead, to do a good job of meeting requirements for leadership, guidance, and command on the part of the party committees and commanders at the various echelons in the sphere of the campaign, to achieve the highest results.

All cryptographic organizations taking part in the campaign ensured service to leadership, guidance, and command of the units by fulfilling the task of preparing for the campaign, taking part in executing the order to open fire and attack the enemy at exactly 1130 hours on 30 March 1972.

The first phase of the campaign, from 30 March 1972 to 1 May 1972, had 159 units with cryptographic organizations, with 487 points of contact among the units: three infantry divisions, two independent regiments and three independent battalions, five battalions of foot sappers and water sappers, artillery units the equivalent of two divisions (not counting divisional artillery regiments), air defense forces (AA, rockets) equivalent to two divisions (not counting the AA battalions accompanying the infantry regiments and the tens of companies of 12.7 and 14.5mm AA of the provincial units from within the North that were participating), two tank regiments, two engineer regiments and the 671st Division, battalions and companies of regional troops, etc. Units from regiment up to division had from two to three CPs and mobile radio stations, besides which there were four rear service nets [cum] of the front serving in various directions.

The second phase of the campaign, from 2 May 1972 to December 1972, saw the situation complicated and additional combat forces, so the number of units with cryptographic organizations increased to 202 and the number of points of contact increased to 614, with additional combat forces participating in phase two comprising three more infantry divisions, an independent regiment, an engineer regiment, the navy's K5, numerous regional battalions and companies from the B4 front, and a number of branch units.

Confronted by the overwhelmingly brave assault by our army, the enemy reacted in a frenzy, concentrating their firepower and aircraft, including B52 strategic aircraft, in fierce attacks continuing day and night. The cryptographic technique system [he thong] expanded and changed continuously. . . there were times such as, once, when personnel and means had to be arranged for from three to four mobile assault stations. There were situations in which urgent contact had to be set up to cover the 48th Regiment (320th Division) during times of temporary subordination to the 304th or 325th divisions, then returned to its line-up with the 320th Division, only to be temporarily resubordinated to a different division. Because of this, arrangements, adjustments, and passing on and receipt of the various types of cryptographic systems were very difficult.

With the line, "neat and light, mobile, continuously and through long days," but still ensuring precisely the principles for the use of technique, the campaign cryptographic organizations zealously surmounted difficulties to organize cryptonets to respond completely to each request for leadership, guidance, and command, on the part of the Main Military Committee, the High Command, the Party Committee and commander of the campaign, as well as those of units in the campaign. Taking the Tri-Thien Campaign, as a example, it marked a step in growth with respect to organization and direction of the cryptographic technique system in a campaign and in large-scale fighting.

The volume of secret messages also exploded with the unfolding development of the campaign, in which cryptography sometimes used technique KTB5 and at others used technique KTC. In primary CPs of the campaign the volume of messages going and coming per day was 340 official messages at the lowest and 664 at the highest. On an average, each person encrypting or decrypting had to handle from 17 to 33 official messages. At division level, the low point was sixty official messages and the highest 100 – the high for a month was 3,008 official messages. At the level of a regiment (artillery) the low was 432 and the high 738 official messages. Nearly all messages during the campaign were of high precedence: Immediate and Priority messages made up 92.88 percent of the total. The Encrypting-Decrypting Bureau of the Cryptographic Directorate had to use sound-powered [tang am] telephones in order to transfer encrypted messages to the campaign CP cryptographic bureaus to ensure routing of the messages speedily and accurately.

The cryptographic organization of the 312th Division, just back from service in the Laotian theater, jumped right into the Tri-Thien Campaign. There were cryptographic comrades disabled while serving command, yet who still sought a means of protecting the security of technical materials. The forward cryptographic organization of the 203rd Brigade (armor) had two comrades, Kieu Xuan Co and Nguyen Van Vinh, who were caught by a B52 strike while encrypting and decrypting and had their underground shelter collapsed. Cde Vinh gave his life; Cde Coi, although disabled, continued to claw in the dirt and gathered together and turned over his materials before consenting to go to the hospital. Many comrades had high fevers but continued to make an effort to ensure the accomplishment of their work. The cryptographic comrades in units engaged in fighting in the district seat of Quang Tri encrypted and decrypted messages under continuous and violent enemy bombardment night and day, and still ensured that operational orders were handled accurately and promptly. Some cryptographic comrades on assignment to outstations encountered enemy ambushes and with their companions fought doggedly: when they got away they still guarded their cryptographic materials and got them safely to the units.

During the entire campaign, forty-four cryptographic comrades and personnel gave their lives for the mission.

Through more than ten months of combat service, Party committees and campaign HQ affirmed that the cryptographic organizations had done a fine job, ensuring that requirements for leadership, guidance, and command were met. Hundreds of organizations and individuals were awarded decorations and commendations.

Along with the cryptographic organizations in the Tri-Thien front, cryptographic organizations in MR 5, the Highlands Front, Eastern Nam Bo, and the Mekong River delta did a good job serving leadership and command in attacking the enemy in those theaters.

In Sector 5 we opened a general offensive campaign aimed at sapping enemy vitality and expanding the liberated region. Our armies and our people uniformly opened fire, attacking hundreds of enemy positions, bases, military subsectors, district seats, airfields, and storage facilities. The cryptographic forces of the MR comprised 500 comrades, ensuring liaison service to more than 800 points, with hundreds of types of technique deployed at three MR CPs – primary CP, northern wing CP, southern wing CP--the 2nd and 3rd divisions, the regiments of artillery, tanks, and antiaircraft,and the regional armed forces. The MR cryptographic organizations performed their mission under conditions of violent bombing and shelling, lacking resources, but continuing to do a good job of serving command in battles such as extermination of the Chu Gan strongpoint, extermination of the De Duc subsector, liberation of the Hoai An and Dong Son district seats, and the hamlet of Tam Quan, exterminating the district seats of Hiep Duc and Hieu Duc, extermination of the cluster of strong points at Cam Doi, those at the Tien Phuoc military subsector, Ba De, the attack on the Mo Duc military sector, Duc Pho, etc.

On the Highlands Front, we opened a campaign to attack, invest, and isolate the enemy. Cryptographic forces of HQ, Highlands Front, the 10th and 320th divisions, regiments, battalions, branches – with more than 200 cadre and personnel – ensured command secrecy in the diversionary plan to draw enemy attention to the north of Dak To and the plan to deploy forces of the units to positions of regroupment. Overcoming difficulties in the tasked area of responsibility in the mountainous jungle region, Highlands Front cryptographic cadre and personnel did a good job of accomplishing their command mission in striking the enemy, totally wiping out hill 1015, hill 1049, attacking the Dak To-Tan Canh defense perimeter, inflicting heavy losses on the puppet 22ndDivision, etc., taking part in the liberation of the northern sector of Kontum province.

In Eastern Nam Bo we launched the "Nguyen Hue" campaign, aimed at eliminating enemy vitality and breaking the lines defending Route 13 and Route 22. This was a drawn-out, combined-branch-operations campaign. The cryptographic organizations taking part in the campaign had to ensure contact for numerous CPs and assault [-unit radio] stations. The cryptographic organizations ensured command secrecy in diverting the enemy in the secondary direction of Route 22, creating surprise for the enemy when we attacked Loc Ninh on 5April 1972. The Cryptographic Bureau of [Southern] Region HQ, the cryptographic organizations of the 203 5th, 7th, and 9th divisions, the 27th Regiment, and those of the branches swiftly expanded to provide timely service to command and to prepare in anticipation of assault requirements, in the process of serving in combat. The Cryptographic Bureau of Region HQ continuously guided the units in taking advantage of training and augmentation, to be able to surge in productivity and quality of service for the campaign.

At the end of May 1972, the Region cryptographic organization and those of Sector 8, the 5th Division,etc., directly served the Region Military Committee and HQ in leadership, guidance, and command in striking the enemy in the general offensive campaign for more than three months, in the sector south and north of Route 4 (My Tho), taking part in wiping out many of the enemy and liberating 35,000 people. Through service in the various campaigns, many cryptographic units and individuals received the appellations, "Outstanding Unit," "Outstanding Individual," responding warmly to the emulation drives to raise the quality and task productivity, put on by the Central Cryptographic Section. The Message Encrypting-Decrypting Section subordinate to the Cryptographic Bureau of HQ, [Southern] Region, received the Order of Liberation Feat of

Arms second class for its performance in serving leadership, guidance, and command in the campaign.

The Nam Bo delta is an area in which getting back and forth is very difficult, with many canals, irrigation ditches, and immense fields of water – it was also a theater of decisive contest between ourselves and the enemy. Although far from the guidance of upper echelon cryptographic organizations and under conditions in which life was difficult, having to be "self-starters and self-suppliers," the cryptographic organizations of MR 8 and MR 9 raised the spirit of self-reliance, surmounting every difficulty to serve leadership and command in attacking the enemy, helping the masses to rise up and destroy oppressors and eliminate shackles, expanding the liberated region. In the process of serving leadership, guidance, and command, many exemplars of courage and sacrifice because of the mission appeared, one example of which was Cde Ba Rang, deputy of the MR cryptographic section. In March 1972, a boat belonging to cryptographic personnel carrying cryptographic materials was crossing the river when it was pounced upon and encircled by eight enemy high-speed boats. The comrades on the boat, dauntlessly and unruffled, guns in hand, prepared to pour bullets into the enemy. But because of having to protect the technical materials, the comrades could not open fire, but yelled out at the enemy to stand off and let our boat go. Faced by the overwhelmingly courageous spirit of our warriors, the enemy panicked and let our boat cross the river. There were also comrades on the road to their assignment who were pursued by a helicopter, and they kept calm, courageously shooting down the helicopter and protecting the security of their technical materials.

The general strategic offensive of our army and our people, with campaigns launched continuously and on large scale in the theaters of the South, Laos, and Cambodia in 1972, struck a deadly blow at the "Vietnamized War" strategy of the American imperialists. Combined with the glorious victory in the defeat of the B52 raids on the North, the Nixon clique had to sign the Paris Accords on Viet Nam on 27 January 1973, pledging to respect the independence, unified sovereignty, and territorial integrity of our nation, terminating American military involvement in Viet Nam.

In the general offensive of 1972 in the Southern theater, the army cryptographic branch performed the mission of ensuring command secrecy with many large-scale, combined-branch-operations campaigns, long in duration and coordinated in timing over the theaters. Through service to command in combat, army cryptography grew up outstandingly in every respect. The ranks of army cryptographic cadre and personnel were put to the test and firmly tempered in combat. The system of cryptographic technique had achieved an advanced level – had the capacity to ensure the requirements of the campaign in every circumstance. The level of guidance and the use of technique were also elevated greatly.

PARTICIPATING IN THE DEFEAT OF AMERICA'S SECOND DESTRUCTIVE WAR AGAINST THE NORTH

The heavy defeats of the Americans and their puppets in 1972 led to the threat of disintegration of the Saigon puppets' lackey army, so the Nixon clique had to mobilize American military forces to return to participate in the war of aggression in Viet Nam. In the South, they increased their air and naval forces participating directly in counterattacks by the puppet army. In the North, on 6 April, Nixon launched the second war of destruction by the air force on economic sectors and large municipalities, sowing mines to blockade our estuaries, ports, and coastal regions.

Confronted by the mad acts of war by the American imperialists, the Central Party Executive Committee decided to continue the strategic offensive in the South, defeat the subversive warfare of the American aggressors, firmly protect the North, and resolutely implement the strategic objectives that had been put forth.

Army cryptographic organizations in the North, from the General Staff's Cryptographic Directorate to the service and branch cryptographic organizations and the basic unit cryptographic organizations, all jumped into the new combat, ensuring service for the tasks of leadership, guidance, and command in beating back the enemy's air forces, serving command in sweeping and destroying mines, striking American warships, and serving leadership, guidance, and command in increasing the ensurance of transportation and aid to the major lines of the South.

By ingenuity and courageous hearts, surmounting every difficulty that tested them, starting with the American imperialists renewing the second destructive war by their air forces, by 27 October 1972 the army and people of the North had brought down 651 aircraft, capturing many pilots alive; had shot into and set afire eighty warships; deactivated and destroyed thousands of mines; beaten the enemy's blockade trickery; and preserved the arteries of transportation and aid for the theaters. On 22 October, Nixon had to announce the cessation of bombing below the 20th parallel.

But with stubborn spirit, through the month of December 1972, the American imperialists launched a strategic assault by B52s against Hanoi and Haiphong on a scale not previously witnessed.

In the stage of beating back the American imperialists' B52s attacking Hanoi and Haiphong, the army cryptographic organizations, especially those of the Air Defense-Air Force, always achieved timeliness for reports on the enemy situation reaching the General Staff, accurately transmitting combat readiness orders down to the units in order to help the command echelons strike back quickly when the B52s arrived to inflict their criminal acts.

In the battle to defeat the second war, the volume of secret and urgent-precedence messages skyrocketed. In the Air Defense-Air Force service, the message volume sent and received during this period was 114,109 secret messages incoming, mainly concentrated on the critical period of twelve days and nights at the end of 1972: Messages so many that they arrived at a level thought impossible to take care of, day and night. The cryptographers worked without knowing fatigue, in order to serve command of the fields of combat. Vis-a-vis the types of secret command messages for strikes against enemy warships and airplanes, the time-factor requirement was extremely tight. There were messages that had to be reckoned in minutes, such as those announcing B52 aircraft activity, announcing enemy gunboats shelling the mainland, announcing targets, the time the enemy would strike, orders mobilizing combat forces, adjusting vehicular formations. . . moving the location of the battlefield, the location for troop stationing, ordering diversions to create conditions to surprise and wipe out enemy aircraft, etc.

During the time of striking back at the B52 aircraft, in order to routinely and continuously grasp the task of serving leadership and command by the Central Military Committee and of HQ, the Encrypting-Decrypting Bureau of the Cryptographic Directorate of the General Staff was expanded into two elements, one to perform the mission at the base sector, the other in Hanoi, to directly serve the Politburo and Central Military Committee. In circumstances in which serving the requirements of command were most urgent – the enemy making violent attacks – the bureau encrypted and decrypted General Staff messages, continuing to ensure that their mission was accomplished well.

On 30 December 1972, the American government had to declare a cessation of bombing from the 20th parallel on up. America's strategic attacks by B52s on Hanoi and Haiphong, and their scheme for hard negotiations on that basis had been ignominiously defeated.

During the twelve days and nights of striking back at the strategic raids by B52s implemented by America,our army and people shot down eighty-one aircraft including thirty-four B52s, and captured alive forty aggressor pilots. Army cryptographic cadre and personnel did their part to achieve the "Dien Bien Phu of the air" of our army and people.

Notes

1. The standard set for use of technique KTB5 met and surpassed the standard for use of technique KTB4. In the Encryption-Decryption Bureau of the Cryptographic Directorate of the General Staff, the highest average productivity in encryption and decryption was 535 groups/hour, topping technique KTB4 by 15 percent. In the Navy, individuals achieved a high average of 746 groups/hour with 100 percent accuracy, equal to 99 percent of the productivity in encrypting and decrypting by means of technique KTB4. The average production of encrypting and decrypting KTB5 for the whole branch in 1970 was 411 groups/hour, surpassing that of KTB4 by 60 groups/hour.

2. Some units achieved an average output of 542 groups/hour; individual high output was 767 to 875 groups/hour with 99.5 percent accuracy; some comrades ensured encryption/decryption of 20,000 groups/month, some 40,000 message groups/month.

3. Extract from the task report of Region military cryptographic for 1969.

4. In 1968 the total number of messages encrypted and decrypted was 448,685 official messages; for the first six months of 1969, it was 431,039 official messages.

Chapter Eight

The Army Cryptographic Branch in the Strategic General Offensive to Liberate the South in the Spring of 1975

The ink had not even dried on the signing of the Paris Accord before it was subverted by the American imperialists and their puppet Saigon regime. In the days following the signing of the Paris Accord, the sounds of gunfire continued to explode over the Southern theater. With American aid, the puppet Saigon regime made every effort to build the puppet army, continued to carry on the war, and launched "flood the territory" operations to pacify and occupy our liberated regions.

Confronted by the schemes and operations of the enemy, the Central Military Committee instructed the armed forces in the South:

> We must seize on the strategy of attack, defeating each of the enemy's pacification/occupation operations, winning the people and keeping the people, holding on to the liberated regions and revolutionary authority. At the same time, we must be ready for every contingency; if the enemy expands the war to provoke larger scale warfare, then we must be determined to exterminate them. [1]

Thoroughly, profoundly, grasping Resolution 21 of the Central Party Executive Committee and the resolution of the Central Military Committee, based upon an accurate appreciation of the mission situation, the cryptographic cadre and personnel clearly determined the political responsibility, and, with revolutionary ardor reinforcing combat will, concentrated on doing their utmost to carry out the specialty missions.

By 1972, because of the complicated situation after the signing of the Paris Accords, the volume of messages which the cryptographic organizations had to take care of continued to be very large. According to incomplete statistics, in the military cryptographic system of the South, for the first six months of 1973, the number of messages sent and received must be reckoned at 632,336 official messages.

In MR 9, the enemy continued to mount operations up to division level, to occupy our liberated regions. Cryptographic organizations did a good job serving MR HQ commanding counterattacks and attacks on the enemy, to hold on to the liberated regions. In March 1973, four comrades from the MR Cryptographic Section continuously decrypted 295 Immediate [toi khan] messages in the KTC technique, containing the Central Military Committee resolution, while the MR committee met, waiting for the contents of this resolution. Afterward, eight comrades encrypted and decrypted continuously 1,460

Priority official messages in two consecutive days, in order to get them promptly to the command comrades.

At the beginning of 1973, a joint [lien hiep] military cryptographic organization was established to serve command leadership and warfare guidance with the enemy in the implementation of the Paris Accords. Four hundred thirty-two army cryptographic cadre and personnel were assembled and stationed in forty cryptographic organizations, comprising the Northern Military Delegation, the Southern Military Delegation B, seven regions, and 30 [control] teams. The liaison network for cryptographic technique was also organized, ensuring thorough grasp from the outset. When they came into contact with the enemy, cryptographic cadre and personnel in the groups and teams comported themselves with the bearing of victors. The comrades were very cool and vigilant in the face of each of the enemy's actions and provocative tricks. There were comrades whom the enemy tried to rob of their technical material pouches, but the comrades had such a determined attitude that the enemy had to back off.

In MR 5, during 1973, the enemy mopped up and occupied on a large scale. The MR cryptographic organization alternated between performing the mission of ensuring service to leadership and combat command in striking the enemy, and in ensuring service to the joint [lien hop] commissions and teams. Joint commissions of the Sector, Region 2, Region 3, the joint teams of nine provinces and cities were in liaison with HQ and the Four-Party Joint Military Commission at Tan Son Nhat. In order to overcome the lack of troop strength, the MR 5 Cryptographic Bureau realigned forces and urgently enrolled students for quick training in order to have personnel to augment the places that were lacking and [still] have forces in reserve. At the same time, the Cryptographic Bureau also organized cadre refresher classes to upgrade their technical level--the volume encrypted and decrypted and sent, using technique KTC – for a number of cryptographic personnel from division and provincial unit levels.

Day and night the Bureau of Encrypting and Decrypting of the General Staff Cryptographic Directorate served to ensure leadership and command from HQ to the theaters of war. The bureau also assigned ten cadre and personnel to go serve the Two-Party and Four-Party Joint Military Commissions and established an encryption-decryption section responsible for the liaison net system for guidance in the implementation of the Paris accord. This was a time in which the message volume the bureau had to handle increased manyfold. The volume of messages to encrypt and decrypt on some days went up to 1,000 official messages. The total number of messages handled by the bureau in 1973 was 205,992 official messages comprising 12,264,222 groups.

In October 1973, I Corps [quan doan] was formed. Along with the formation of the corps, the cryptographic organization took shape, comprising the corps Cryptographic Bureau, the Cryptographic Sections of the 308th, 312th, 320th, and 367th divisions, the cryptographic organizations of the 45th, 202nd, and 299th brigades, the cryptographic teams at battalion level, etc. Comrade Nguyen Quoc Sung was appointed chief of the corps cryptographic bureau. The Cryptographic Directorate of the General Staff guided and assisted the I Corps cryptographic organization to quickly settle its organization; urgently get into implementation, as an orderly routine, of the tasks of serving to ensure command; training; and the other aspects of the professional task, in order to do a good job from the outset of implementing the mission of the cryptographic organization in the first corps of our army. Also in 1973 the system of cryptographic organization expanded into the MRs, services, and branches, i.e., establishing the Cryptographic Section of the 919th Air Force Transport Brigade, the Cryptographic Section of the 673rd Air Defense Division, etc.

Working while building, in every aspect, cryptographic organizations army-wide stepped up their study and raised productivity and quality in the use of technique KTC. By the end of 1973, the cryptographic organizations in MR Viet Bac, MR Northwest, MR Left Bank, MR IV, the Naval Service, a number of divisions belonging to the Air Defense-Air Force Service, and Armor HQ had stopped using KTB5 and changed over to technique KTC5. The cryptographic organizations of the 320B Division, 304th Division, and 320th Division began to expand the study and use of KTC. Cryptographic organizations in MR 5, the Highlands Front, and in Nam Bo also expanded the training and use of technique KTC down to basic units, with a sense of immediacy and zealousness.

The Cryptographic Directorate of the General Staff summarized and disseminated the concrete experiences concerning the tasks of organization, training, and use of KTC, especially the experience of training in the basic technical subjects of encrypting and decrypting, helping cryptographic organizations at various levels army-wide to achieve good results.

In the movement to study the use of KTC bubbling through the entire army, there were many units and individuals who achieved rather high productivity and volume. The Encrypting-Decrypting Bureau of the General Staff Cryptographic Directorate achieved an average productivity in encrypting and decrypting of 542 groups per hour, with a 99.90 percent accuracy, and many comrades achieved record highs, such as Cde [Miss] Pham Thi Muon, with 750 groups per hour, Cde [Miss] Pham Thi Vien, with 740 groups per hour, etc.

During this time, many cryptographic cadre and personnel from the Southern theater came to the North for treatment of disease, convalescence, and study. On such occasions, the Cryptographic Directorate organized refresher courses in techniques and · professionalism. Cryptographic organizations of the MRs, services, and branches also proceeded to organize training in the various task aspects for cadre and personnel. The Cryptographic School stepped up the training of new personnel to understand the use of both types of technique (KTB5 and KTC) in order to augment the essential theaters, principally the Southern theater.

Carrying out the instructions from the Central Cryptographic Section, the army cryptographic branch implemented a summarization of the cryptographic task in eight years of serving the resistance against America. Vis-a-vis this task, the army cryptographic organization executed it step by step, constantly and continuously, throughout the period of opposing America and saving the nation. Thanks to doing a good job of recapitulation, the army cryptographic branch promptly drew experience, publicized achievements and strong points, while, at the same time, quickly resolving shortcomings and promptly taking corrective action to get every aspect into the routine.

In 1969, after a recapitulation of the task, the MR Left Bank Cryptographic Bureau wrote two documents, "Raising Productivity and Quality of Encrypting and Decrypting to Serve Victorious Combat," and "The Independent Cryptographic Task."

Through service to the campaigns to strike the enemy and ensure lines of communication and transportation in MR 4 and the 1972 strategic general offensive campaign (for example, the task of cryptographic service in the Tri-Thien campaign and the task of serving command leadership of [Southern] Region HQ), through the periods and campaigns of defeating the American imperialists' strategic assaults by B52s on Hanoi and Haiphong in December 1972, the bureaus of cryptography, the Encrypting-Decrypting Bureau of the General Staff Directorate of Cryptography, the [Southern] Region Cryptographic Bureau, cryptographic bureaus of the MRs, services and branches-- all recapitulated and drew experience concerning the thorough grasp of mission, concerning the task of organizing technical networks [he tong], implementing encryption and decryption, and ensuring the flow of outgoing and incoming messages.

During the period of resisting America's war of destruction, the Cryptographic Directorate of the General Staff collected and sorted out situations, assessments, appreciations of accomplishments, good points and bad points in implementation of the missions of the various cryptographic organizations, and extracted major experiences of value to professional leadership for the whole branch. Based upon summarization of the documentation and of the experiences, the General Staff Cryptographic Directorate compiled and produced documents that put forth reasoning and technical professional practices to nourish and elevate the specialty level of cadre and personnel, e.g., "Organization and Implementation of the Cryptographic Task of the People's Army of Viet Nam in War Time," "Message Error Detection and Prediction," "Methods of Training in the Four Primary Technical Subjects and Raising the Productivity of Double Encryption," and "The Independent Cryptographic Task."

Summarizing the eight-year national salvation struggle against America (1965-1972), the accomplishments of the army cryptographic branch that stand out greatly are having organized and done a good job of executing the task of encrypting and decrypting messages, ensuring that 50,008,006 secret messages, sent and received, were secure, accurate, and timely, not allowing error to influence the tasks of leadership, direction, and command, especially having implemented in outstanding fashion the policy of changeover to new technique under circumstances in which the task was difficult, complicated, with endless hardship and fierce fighting. After eight years of being put to the test, the army cryptographic branch had built the ranks of cadre and personnel to 5,337 comrades of good political quality, steadfast and tempered, with high revolutionary ardor, with a sense of responsibility for the political mission of the branch, and with a sense of responsibility for organization and discipline – a spirit of overcoming obstacles and withstanding hardships – sacrifice of life – diligently studying to raise the level of ability – to accomplish the mission--to be worthy of being members of the Lao Dong Party of Viet Nam. Some 450 comrades had given their lives for their country.

In March 1974, a conference was convened in Hanoi to recapitulate the cryptographic mission of eight years of service in the national salvation struggle against America (1965–1972) on the part of Viet Nam's cryptographic branch. Groups of cryptographic delegates from the networks of Party and government, army and public security, representatives of cryptographic organizations from the entire nation, and cryptographic teams on international duty came back to participate adequately. The delegation of the army cryptographic branch, led by the comrade chief of the directorate and comprising comrade representatives of the General Staff Directorate of Cryptography and representatives of the MRs, services, branches, and the unit organizations, arrived to participate in the conference.

The conference took place ebullient in the impetus of victory, with a sense of confidence and heightened unanimity. This was a conference of most important significance, for it marked the growing up of the Vietnamese cryptographic branch in the process of combat, building and expanding, in the task of ensuring leadership and command of the revolutionary war.

The conference was graced with the presence of Cde Le Duc Tho, member of the Politburo and Secretary of the Central Party and Cde Nguyen Don, member of the Central Party and Deputy Chief of the General Staff,who visited and spoke. Cde Le Duc Tho commended the accomplishments of the cryptographic branch: "We have come through eighteen years of resistance to America, saving the nation (1955-1972), and have achieved a great victory. Comrades still living as well as those who have given their lives – all have made a worthy contribution to this great victory of our race . . . Today I come to speak to you all in order to express the sentiments of the Central Party toward the cryptographic branch, toward you comrades, and also to commend you comrades who have made a worthy contribution yourselves toward the work of the Party, the work of our race in resisting America and saving the nation." After speaking clearly of the position, the role, and the concrete accomplishments of the cryptographic task, Cde Le Duc Tho continued with feeling: "Central is very pleased and very moved that you comrades have made great efforts, have given your lives. Nearly 500 cadre and personnel have fallen, a proportion that was high, for this was many, not few; this sacrifice was not less than that of soldiers on the field of battle; this sacrifice was rather large. Here, too, was a front with large casualties."[2]

The comrade thoughtfully suggested: "The victory road of revolution is indispensable – we shall absolutely liberate the South and unify the nation, but if we wish to achieve victory, we still must pass through steps on a hard and difficult road. The mission of you comrades is still very exacting – you still must plow a long way through the theaters of war, you must transmit the instructions and resolutions of the Party, and of the government, from organizations at the nerve center out to places on all three fronts – military, political, and foreign affairs. . . You comrades must study to raise the level of scientific professional technique, to raise the level – more modern, more creative. When you conclude the conference, people in this front – people in that front, people in the North, people in the South, people in foreign nations, spread out everywhere in theaters and fronts to perform the mission, although saying not a word, are quite glorious."

Cde Nguyen Don, on behalf of the Central Military Committee, congratulated the cryptographic branch for having performed its mission in an outstanding manner, through eighteen years of resistance to America and saving the nation. He analyzed deeply and concretely the achievements, good and bad points, and the reasons for these good and bad points from the organizational and technical aspects. He also clearly indicated to the cryptographic branch delegates the way for the branch to strive to advance. The delegates were moved and enthused beyond measure at the words of congratulation and counsel from the leadership comrades representing the Central Executive Committee, promising the Central Party Executive Committee and Central Military Committee that they would mobilize the cadre and personnel to do a good job of carrying out the comrades' instructions.

The conference recapitulating the cryptographic task in eight years of resistance to America and saving the nation came off beautifully. From the atmosphere of the conference was created a powerful, ebullient emulation movement in the task for the whole cryptographic branch in general and the army cryptographic branch in particular.

In May 1974, II Corps was formed, the cryptographic organization comprising the II Corps Cryptographic Bureau, the cryptographic sections of the 304th, 324th, 325th, and 673rd divisions, and the cryptographic organizations of the 219th, 203rd, and 164th brigades, with comrade Le Ngoc Luong chief of the Corps Cryptographic Bureau. Immediately upon the formation of the corps, the corps system of cryptography was able to ensure command service during the campaigns to liberate the district capital of Thuong Duc (Sector 5), the K18 campaign (Hue), and to serve command in preparing for battle and in the fighting against the enemy, to hold onto the Quang Tri liberated region.

In July 1974, IV Corps was formed in the Eastern Nam Bo base region. The cryptographic organization of the corps took shape, comprising the cryptographic sections of the 7th and 9th divisions and the cryptographic organizations of the regiments and branch units.

During this time, cryptographic organizations throughout the army did a good job of accomplishing the task of encrypting and decrypting messages, serving the work of thoroughly grasping leadership and precisely implementing Central Party Resolution 21 of the Central Party and resolutions of the Central Military Committee, guiding, shaping, and molding the theaters and regions in holding fast to the viewpoint and revolutionary ideology of attack, serving to guide the building and completion of the system of strategic and campaign lines and serving command in the transportation and supply of the theaters.

The cryptographic organization in Eastern Nam Bo served command in the liberation of Phuoc Long, the first province in Nam Bo to be liberated. During this time, the Central Cryptographic Section and the Cryptographic Directorate of the General Staff zealously instructed the implementation of enrollment and development of new personnel, research into the production of various types of dictionary codes, cryptographic key, and command opcodes, according to the norm of the 1974–1975 two-year plan (which anticipated the development of 1,000 personnel, research and production of 1,500 types of dictionary codes, 3,000 sets of cryptographic key and 1,500 command opcodes). From directorate head to professional organizations, there was an increase in going to inspect and assist on the spot the cryptographic organizations throughout the army, from North to South and Laos.

THE ARMY CRYPTOGRAPHIC BRANCH IN THE GENERAL OFFENSIVE AND UPRISING OF SPRING 1975 AND THE HISTORIC HO CHI MINH CAMPAIGN

From 18 December 1974 to 8 January 1975, the Politburo of Central Party met. After examining and analyzing the unfolding situation from every aspect, the Politburo resolutely determined the following strategy:

All-out mobilization of the power of the entire Party, the entire military, the entire people, in both of the two areas in 1975–1976, stepping up the military and political struggle, combined with the diplomatic struggle, to cause rapid change and across-the-board force comparison in the Southern area theater advantageous to us, implementing as a matter of urgency and accomplishing each preparatory task, creating conditions ripe for general assault, general uprising, to eliminate the puppet army and cause it to disintegrate, to strike and bring down the puppet authorities from central to regional, placing authority back in the hands of the people, and liberating the Southern area. If the opportunity arose at the beginning or end of 1975, then immediately to liberate the South in 1975.

On 10 January 1975, executing orders from the General Staff, the Cryptographic Directorate of the General Staff organized a cryptographic team of seven comrades[3] to serve Group A75 under Gen. Van Tien Dzung, chief of the General Staff, secretly going down to the Highlands in order to research, organize, and realize the Politburo's strategic decision.

The Encrypting-Decrypting Bureau of the Cryptographic Directorate of the General Staff made as its first priority the encrypting and decrypting of Group A75 messages in order to regularly serve communiques of the latest news, especially information concerning the unfolding enemy situation and the transportation situation to ensure the campaign, so that HQ could make an operations plan.

The army cryptographic branch thoroughly and profoundly grasped the important change in the revolutionary mission situation. Cryptographic organizations throughout the army enthusiastically, cheerfully, and with confidence proceeded to make preparation in every respect, so as to be ready to receive and execute the missions received from the Central Military Committee, HQ and commissars, and commanders at the various echelons.

The General Staff Cryptographic Directorate also – along with cryptographic organizations of the MRs, services, branches, etc. – researched and made concrete plans and implemented the task of supplementing, correcting, and aligning the cadre and personnel for the units, principally the main point units, with special urgency for the strategic mobile units and the key theaters.

The Army Cryptographic School picked student comrades who were coming along well and were clever, and organized them into individually assembled companies, and brought them up to speed technically and professionally in a short time, so as to have strategic reserve forces and be able to quickly assist the theaters. With an all-out effort, and after a short time, the school had prepared more than 300 comrades, ready to set out and perform the mission.

The cryptographic organizations of MR Viet Bac, MR Left Bank, MR Right Bank, etc., selected comrades of good technical ability and qualities so that, when the orders came, they could supplement and augment the forward units.

In lock step with the preparation and expansion of the organizational task, the Cryptographic Directorate issued instructions to expand the systems [he tong] and means of cryptographic technique, urgently getting off a large volume of the various types of technique and professional means and equipment for the theaters. With an outstanding all-out effort, by 1 March 1975, before our army had opened fire to raise the curtain on the Spring 1975 General Offensive, cryptographic forces had expanded all over the Southern area theater as follows:

On the Highlands Front, our army opened the assault campaign under the name, "The 275 Campaign," the number of cryptographic cadre and personnel participating being 456 comrades, ensuring liaison for 374 points with seventy types of technique, the responsibility of the cryptographic organizations being to serve leadership and command of the campaign CP, the divisions, the branch units, and the main force and regional units in the Highlands area.

On the Tri-Thien-Hue Front (Front B), afterwards called the 475 Front, the cryptographic cadre and personnel resources comprised 1,144 comrades, ensuring liaison for 1,241 points, with 195 types of technique in use, the responsibility of the Tri-Thien-Hue Front cryptographic being to serve leadership and command of the Forward HQ CPs, the primary CPs, the Forward CPs of MR Tri-Thien, MR V, II Corps, the 2nd, 3rd, 324th, 325th, and 304th divisions, the 52nd Brigade, and the branch units in the area, plus the regional units.

On the Southwest and South Saigon Front, the cryptographic cadre and personnel resources comprised 1,004 comrades along with 677 comrades of subordinate units, ensuring liaison for 1,678 points with 154 types of cryptographic technique in use. The responsibility of the cryptographic organizations of the Southwest and South Saigon Front was to serve leadership and command of Southern Area HQ; MRs 7, 8, 9; IV Corps; the branch units; the main force units in the area; and the regional units.

The cryptographic organizations of Air Defense-Air Force and Navy services and the Sapper, Armor, Engineer, and Artillery branches also developed with respect to organization and technique both widely and deeply into the Southern theaters in order to serve the duties of leadership and operational command of combined services. The Navy cryptographic network developed its campaign service duty to comprise 125 liaison points, internally, skip-echelon, direct and combined [or joint, operations]. The Air Defense cryptographic network developed at the peak time eighty-two stations, 202 points, including three forward CPs.

The cryptographic organizations in the entire army in the rear were all in the position of readiness with support people and professional technical means for the cryptographic organizations up front, in order to ensure accomplishment of the mission of the General Offensive Campaign of Spring 1975.

The total number of army cryptographic forces taking part directly in the service of the general strategic offensive and the historic Ho Chi Minh Campaign was 4,167 cadre and personnel, ensuring as a system a liaison net that, at its largest, was 3,703 places, using 419 types of cryptographic technique. Cryptographic organizations of the campaign were equipped with an additional tens of tons of professional technical means. It can be said that this was an outstanding effort on the part of the army cryptographic branch, which had never before planned to organize to ensure command leadership and operational guidance that was adequately prepared and sufficient to cover every aspect of this spring of 1975.

On 10 March 1975, the Spring 1975 general offensive and uprising opened with the daring surprise strike on the town of Ban Me Thuot. The cryptographic organizations participated in totally ensuring secrecy as to the objective of the campaign and ensuring the command secrecy of the diversionary operation of Campaign HQ (e.g., the diversionary operation of the 968th Division) to draw enemy attention to the north of the Highlands and create surprise for them when our army opened fire and struck the town of Ban Me Thuot. Dummy messages – deception messages – were continuously sent into the air on a daily basis. Military information was speedily sent back to HQ. "Every tiny movement in the theater was closely followed by the operations watch – every forward step by the troops was quickly marked on the map. The fighting in the Highlands, with Ban Me Thuot at the center, was at this time the number one concern of the Politburo and Central Military Committee – and those in charge at the General Staff and the directorates."[4]

When we victoriously assaulted the town of Ban Me Thuot, the cryptographic organization quickly passed on the contents of messages from the Politburo, Central Military Committee, and High Command -messages for the Highlands Front, encouraging and commending the cadre and warriors, directly guiding the expansion of the campaign with the spirit of "creativity, daring, and urgency," to grasp the opportunity to secure an even greater victory. At the same time, the cryptographic organization also passed along the contents of electrical communiques for Sector 5, B2, and Tri-Thien concerning the guidelines of the upper echelons following the Ban Me Thuot victory. The Highlands Front cryptographic organization served to ensure combat command in shattering the enemy counterattack aimed at retaking Ban Me Thuot, and served to ensure command in pursuit and interception of the enemy when they fled, abandoning the Highlands and took part in exterminating and dispersing the puppet II Corps, liberating the entirety of the Highlands.

On 18 March 1975, the Politburo of the Central Party Executive Committee met. Through analyzing and estimating the situation with respect to the victory of strategic significance for our side, the Politburo and Central Military Committee unanimously resolved to produce a plan for the liberation of the Southern area in 1975, determining that the main strategic direction of attack would be Saigon, prior to which would be the extermination of all enemy forces in their MR 1 and the liberation of Hue, Da Nang, and the provinces of Central Viet Nam.

Implementing instructions from the Politburo and Central Military Committee, the General Staff prepared at once a plan for a large-scale attack to wipe out the forces in the enemy's MR 1, and to liberate Hue and Da Nang.

The General Staff cryptographic organization swiftly and precisely sent operations orders from the High Command to II Corps, MR Tri-Thien, and MR 5. "Paying no mind whether day or night, communications and cryptographic personnel assigned to duty in Sector A received and decrypted messages in a spirit of highest urgency, requiring the greatest precision, in order to keep the operations watch comrades posted, whatever the hour. The fellows and girls of communications and cryptography were educated and imbued with the degree of importance of each individual, each element, in these days and months of urgency and consecutive victories." Cryptographic organizations of II Corps and MR Tri-Thien directly passed on the command orders of the General Staff [Bo] for II Corps and MR Tri-Thien to clearly receive opportunities, grasp deep-thrust targets, and isolate, surround, and interdict the enemy's lines of withdrawal.

On 25 March 1975, Hue City and Thua Thien province were liberated.

On 25 March 1975, the Quang Da Front Command [Bo chi huy] was established under Cde Le Trong Tanas commander, Cde Chu Huy Man as political commissar, with the front cryptographic organization comprising seventy-two cadre and personnel under Cde Tran Ha.

Cryptographic organizations from the Encryption-Decryption Bureau of the General Staff down to the front cryptographic organization, and those of II Corps and MR 5 speedily set up a cryptographic technique system to serve the leadership and command guidance of HQ.

Immediately upon receiving the message of instruction from the Politburo and Central Military Committee (dated 24 March 1975, sent to MR 5 and II Corps, concerning guidelines planned for the liberation of Da Nang), the Cryptographic Bureau of MR 5 received orders from the MR to quickly develop a liaison net for HQ with direct subordinates, HQ forward, [Southern] Region Military HQ, the MR primary CP, lateral communication with the divisions of II Corps, internal liaison with the branch regiments, the primary CPs, Quang Da Forward, etc.

"Messages from the General Staff [Bo] to the Tri-Thien and Quang Da fronts were continuous, thick and furious, during these days, exuding requirements for all-out urgency, for swiftly pressing and surrounding and dispersing and wiping out the enemy in this second strategic pummeling."

The cryptographic organizations of MR 5 and II Corps promptly handled orders from the High Command and served HQ, II Corps and HQ MR 5 commanding a swift spreading out to execute the assault on Da Nang. On 29 March 1975, Da Nang--the second largest city in the South--was liberated. We eliminated and scattered a large force of the enemy army, and broke through the new strategic defensive system of the enemy.

The Hue-Da Nang victory was of great significance, along with the Highlands victory, in changing for good the balance of forces between ourselves and the enemy, creating favorable conditions for us to launch the campaign to liberate Saigon.

After Da Nang was liberated, at 1630 hours 29 March, the General Staff cryptographic organization encrypted a message from Cde Le Dzuan to the theater:

> The situation is changing rapidly – the revolution in the South is entering the stage of spreading by leaps and bounds. I concur with you fellows that this is the time in which we need to act promptly in an all-out effort, determined and daring. In reality, it may be considered that the campaign to liberate Saigon began at this point .... Good health and great victory to you all.

On 31 March 1975, the Central Party Politburo convened under Cde Executive Secretary Le Dzuan. In this meeting the Politburo affirmed: the revolutionary struggle in the South not only had entered the stage of spreading by leaps and bounds, but also the strategic opportunity to carry out the general offensive and uprising was ripe. Thus the Politburo resolved:

"Grasping the strategic opportunity more than before, with ideological guidance – like lightning, daring,with the element of surprise, and certain victory – be determined to carry out the general offensive and uprising in the nearest timeframe, preferably April, without delay."

A week before this historic session, the Politburo appointed Cde Le Duc Tho, member of the Central Party Politburo, to go into the South, so that, along with comrades Pham Hung and Van Tien Dzung, guidance for the offensive and uprising could be provided. Cde Nguyen Van Thinh and Cde Tran Diep were the cadre and individual from the Bureau of Encrypting-Decrypting of the General Staff selected to go serve the transfer of Cde Le Duc Tho for this task.

In order to speedily concentrate forces to win victory in the final battle, the main force corps – III Corps,* II Corps, and I Corps – received orders for speedy movement down to the Saigon-Gia Dinh front.

On 7 April 1975, the General Staff cryptographic organization sent Immediate message #157 from General Vo Nguyen Giap to the units: HQ, Group 559; 559th Forward; the corps; and the services and branches on the march, the entire text being as follows:

"1. Like lightning, and even more so; recklessly, and even more so; taking advantage of every hour, every minute, rush to the front and liberate the South. Be resolved to fight and totally win.

"2. Transmit at once to the party members and soldiers.

VAN [Vo Nguyen Giap]"**

In these days the army cryptographic organization from General Staff Cryptographic Directorate down to unit cryptographic organizations ensured good service to the leadership of the Politburo, the Central Military Committee, and the High Command [extended] to the theaters and regions.

At each step of the march of the wings of the army, their victories from the theaters came flooding back in messages. The atmosphere of work in the Bureau of Encrypting-Decrypting 224 in the Directorate of Cryptography of the General Staff during these days was recalled by one comrade cryptographer as follows:

> The volume of work increased, fast and furious. We spread out our material to work at once. The pace of encryption and decryption increased without let-up: seven minutes, six minutes, five minutes, even four and a half minutes a message. Knocking off a minute was extremely valuable at this time.[5]

Many high-precedence messages went to the wings of the army, overseeing and urging on the axes to hurry up more, to speed the advance. In the wing to the east, when they received a message, cryptographic took it up to Cde Le Trong Tan, who joyfully embraced the comrade cryptographer, Vu Van Canh, and wrote on the message form "hoan ho co yeu thong tin rat kip thoi" ["hurrah for the cryptographers and commo – very timely!"].

---

* III Corps was formed 27 March 1975 in the Highlands, comprising the 10th, 316th, and 320A infantry divisions, 675th artillery regiment, 312th air defense, 198th sapper, 273rd tank, 545th engineer, and 29th communications regiments. Commander, Brig. Gen. Vu Lang; Sen. Col. Dang Vu Hiep, political commissar. *The History of the People's Army of Viet Nam* (Hanoi: Institute for Vietnamese Military History, 1990), Vol. II, Part 2, 247, 247n, 248. – Tr./Ed.

**Ibid., 258, identifies this historic message as #157-H-TK sent at 0930 hours, but adds that copies were sent also to the component divisions and to General Le Trong Tan. --Tr./Ed.]

On 8 April 1975, the Politburo decided to establish the Campaign CP for the liberation of Saigon-Gia Dinh, with Cde Van Tien Dzung as commander and Cde Pham Hung as political commissar, the B2 Cryptographic Bureau being shifted over to perform the mission of campaign cryptographic bureau.

On 14 April 1975, the Politburo and the Central Military Committee approved the plan for the liberation of Saigon-Gia Dinh. The cryptographic organization sent secret message #37 from the Politburo to the campaign CP. At 1900 hrs the same day, the campaign CP received it and among its contents was:

"We agree that the Saigon campaign be called the Ho Chi Minh Campaign."

In the days that followed, many secret messages of the greatest importance from the Politburo and the Main Military Committee were taken care of by the cryptographic organization, dispatching them to the theaters at top speed and with total accuracy. For the army cryptographic warriors, these were hours and minutes of the happiest flapping, honored to be handling the offensive orders of the Party – of the nation – in a period of historic importance for our race.

At 1400 hrs on 15 April 1975, at the Supreme Command Post, General Vo Nguyen Giap handed the mission to Cde Nguyen Dzuy Phe, Director of the Army Cryptographic Directorate. Cde Vo Nguyen Giap instructed: "In the recent days of special and urgent combat by our military and people on the Southern front, cryptographic cadre, soldiers, and personnel accomplished their mission in an outstanding way. The Central Military Committee commends all comrades. The combat that is under way and near at hand until total victory is urgent and decisive. The mission of ensuring the secrecy, accuracy, and timeliness of the content of orders, guidance, and commands from the Politburo, the Central Military Committee, and the High Command will be decisive vis-a-vis our determination to accomplish the liberation of the South. All comrade cadre and soldiers, Party members, group members, and personnel of the Cryptographic Directorate must be highly resolved to seek every means of ensuring this requirement."[6]

Implementing the instructions of the Comrade General, the chief of the Cryptographic Directorate encouraged the entire organization to a higher level of political responsibility, to strive upward in accomplishing the mission that had been entrusted.

In order to promptly handle important messages of the Politburo, the Central Military Committee, and the High Command going to the steering comrades in the theaters, a cryptographic team under Cde Vo Minh Chau, comprising Nguyen Xuan Phu, [Miss] Dang Thi Muon, [Miss] Vu Thi Trong, Nguyen Van Khoi, et al., was sent up to encrypt and decrypt messages right on the spot, in the work place of the Central Military Committee. In the duty team encrypting and decrypting at the work place of the Military Committee, there was very close coordination, quick reaction, and creativity with the radio team, so that messages could get out at once and the fastest encryption and decryption could be ensured. The Comrade General, through personal association and good cheer, encouraged the comrade cadre and personnel in the teams to be calm and self-confident, demonstrating speed and accuracy for the combat orders and communiques of the various theaters. In the room where they worked, cryptographic cadre and personnel encrypted and decrypted extremely important and most urgent messages carrying the signatures of comrades BA (Le Dzuan), TRUONG CHINH, TO (Pham Van Dong), VAN (Vo Nguyen Giap), THANH (Hoang Van Thai) . . . and comrades SAU THO, TUAN, BAY CUONG[7]. . . messages written mainly by Cde Le Dzuan. Ordinarily General Vo Nguyen Giap personally handed messages to cryptographic: having any section of a message written by the comrade general, cryptographic encrypted that section and conveyed it at once to Communications-- there were times when the comrade sat down in the room where cryptographic was at work in order to write or correct messages. Having an incoming message, once decrypted cryptographic sent it along to the leadership comrades present. With a very long message, but a requirement to communicate it most urgently, every minute, every hour counted (such as the message Cde Le Duc Tho sent back on 25 April 1975, a ten-page typewritten communique concerning the situation in theater B2: it amounted to a message of fifteen to twenty pages).

One day around the end of April, after hearing Cde Nguyen Dzuy Phe report on the situation of the Cryptographic Directorate's service to the campaign command and control, Cde Hoang Van Thai arrived to visit and encourage the young men and women cryptographic cadre and personnel on duty serving the campaign in the work place of the Central Military Committee. Cde Hoang Van Thai praised the accomplishments of the Cryptographic Directorate's independent-activity cryptographic teams serving the quadripartite military mission and the groups of Cdes Van Tien Dzung, Le Duc Tho, and Le Trong Tan, and of the duty cryptographic team at the work place of the Central Military Committee. He issued instructions and requested that service to steerage and command in the upcoming campaign be performed such that the Politburo and the Military Committee be able to promptly grasp each stage of development of the wings of the army, and of each point of attack into the last lair of the enemy.

In these historic days of the spring of 1975, on every part of the national soil, life was motivated by the highest magnanimity. All as one people bravely advanced to achieve the final victory. The military forces blitzed into the liberation of Saigon, cryptographic cadre and personnel (616 comrades just out of school and 192 comrades in the units) rushing along as comrades in arms to the front, liberating the entire South.

The cryptographic organization of I Corps alternately engaged in operations and in the ensurance of orders transmitted by HQ and Corps headquarters, commanding the mechanized blitzkrieg troops advancing secretly along a 1,700 km stretch into the Dong Xoai sector (eastern part of Nam Bo) and taking up a consolidated position promptly in accordance with instructions from HQ.

The cryptographic organization of II Corps, after accomplishing the mission of ensuring command in the operation to liberate Hue-Da Nang, also as a unit alternated in operations and ensuring the corps command of the troops "striking the enemy and moving on, opening the road and advancing" from Da Nang down to Bien Hoa,Ba Ria, etc., on the line of advance right up to the gates on the eastern side of Saigon.

The cryptographic organization of III Corps, after serving command in the operation to liberate the Highlands and the southern provinces of Sector 5, as a matter of urgency supplemented and readjusted cadre and personnel and the types of technique to serve the corps commanding assault troops, opening the route of advancement to regroupment positions controlling the jump-off point for the assault on Saigon from the northwest, in accord with the plan of the campaign CP.

The cryptographic organization of IV Corps, after serving the corps HQ commanding the liberation of the town of Xuan Loc, continued to serve the corps command appointing units to stick close to Saigon and prepare to serve the corps assault.

The cryptographic organization of Group 232[8] swiftly developed an organizational system and a technical system, preparing to serve the command plan for the Saigon assault from the south and southwest.

The cryptographic organizations of the 5th, 3rd, and 9th divisions, and the cryptographic organizations of MR 8 and MR 9 served command and control of the mobile forces, isolating Saigon from the Mekong delta, wiping out subsector military posts of the enemy, liberating hamlets, etc., creating a springboard for attacking Saigon from the south and southwest.

The cryptographic organization of MR 5, after serving command and control in the operation to liberate Hue-Da Nang, continued to serve HQ commanding the units which, in turn, were wiping out the enemy's defensive system to the south of the military region, and to serve the mobile operational command without interruption, along a line of nearly 500 km from Quang Ngai to Nha Trang and Cam Ranh, etc., having occasions of quickly setting up liaison points – the MR Cryptographic Bureau had to swiftly transfer cryptographic key from the primary CP of the MR to the 3rd Division, some 300 km distant, to enable the 3rd Division to be in prompt touch with HQ.

While the various wings of the army were closing in on Saigon, MR 5 Cryptographic and cryptographic of the Navy units served command in liberating the Spratly Islands after an operation of three days and nights braving waves and wind.

The cryptographic liaison net system of the Navy in turn was arranged on the naval bases and ports in the South of our nation.

On 22 April 1975, the Encrypting-Decrypting Bureau of the Cryptographic Directorate of the General Staff handled a telegram from the Politburo, signed by Cde General Secretary Le Dzuan, and addressed to the campaign CP:

> The military and political opportunity to open the assault on Saigon has ripened. We need to take advantage of each day--to promptly mobilize the attack on the enemy from every direction, without ceasing . . . you are all to issue instructions at once to the various directions to act promptly. . . .

At precisely 1700 hrs on 26 April 1975, our artillery opened up on the puppet Armor School, opening the curtain for our final assault on Saigon. Cryptographic of II Corps handled the command order to wipe out some important positions, among them the Nuoc Trong base, in order to establish a springboard to facilitate the field of fire of the 130mm guns laying on Tan Son Nhat airfield.

The cryptographic organization of I Corps served the combat command of the corps in the northern and northeastern approaches.

The cryptographic organization of II Corps served the combat command of the corps in the southeastern approach.

The cryptographic organization of III Corps served the combat command of the corps in the northwestern approach.

The cryptographic organization of IV Corps served the combat command of the corps in the eastern approach.

The cryptographic organization of Group 232 served the combat command of the group on the western and southwestern approaches.

On 28 April 1975, while the cryptographic organizations of the five wings of the army were ensuring the service of combat command liberating Saigon, the cryptographic organization of the Air Force Forward CP and the cryptographic team of the hastily established station went along to serve the comrade commander of the Air Force ensuring command of the flight of A37s (aircraft taken from the enemy) in charge of Nguyen Thanh Trung, suddenly bombing Tan Son Nhat airfield.

At 0500 on the morning of 29 April 1975 the wings of our army simultaneously opened fire and assaulted the capital.

Cryptographic cadre and personnel speedily handled messages from the Politburo and the Central Military Committee mobilizing all cadre and soldiers with great fortitude to win total victory in the historic Ho Chi Minh campaign. In that connection, the General Staff cryptographic organization ensured continuous handling of directive messages from the Politburo, the Central Military Committee, and the High Command, sent to the campaign command post.

The cryptographic organization of I Corps served the corps command eliminating enemy bases, hitting and occupying the puppet General Staff [compound].

The cryptographic organization of III Corps served the corps command hitting and occupying Tan Son Nhut airfield, afterward serving command coordination with I Corps, hitting and occupying the puppet General Staff [compound] and advancing toward Independence Square.

The cryptographic organization of Group 232 served command hitting and occupying the Capital Special Sector HQ and the Main Police HQ, the Navy HQ, and the Nha Be gasoline depot.

The cryptographic organization of II Corps served command hitting the enemy and opening the route for advancing on and occupying Independence Square.

The cryptographic organizations of IV Corps and the 3rd Division (MR 5) served corps command hitting and occupying the CPs of MR 3, the Bien Hoa Military Sector, the Thu Duc Special Forces HQ, etc.

At 1130 hrs on 30 April 1975, the liberation banner was unfurled over a housetop at Independence Square, puppet president Dzuong Van Minh having to announce unconditional surrender. The cryptographic team of a II Corps penetration unit (consisting of Cdes Than and Vong) had the honor of being present in Independence Square at this historic hour and minute and received from higher echelons the responsibility of safeguarding the official seal of the puppet Saigon authorities.

The historic Ho Chi Minh campaign was totally victorious. Cde Nguyen Dzuy Phe, Director of the Army Cryptographic Directorate, carried up to the Politburo and the Central Military Committee the telegram announcing that our forces had raised the flag over Independence Square.

In the fifty-five days and nights of the general offensive campaign and the spring uprising of 1975, army cryptographic had ensured the encrypting and decrypting for transmittal of 1,192,525 telegrams, ensuring secrecy, accuracy and timeliness to satisfy the requirements of a blitz advance and the enormously large-scale operation of our army and our people. The military cryptographic of the South alone had handled the encrypting, decrypting, and transmittal of 810,387 official messages. From the General Staff Cryptographic Directorate to the cryptographic organizations of the various levels, there had been applied a method of creating highly valuable experiences, summarized through thirty years of building and fighting, principally in the operations of Khe Sanh and Route 9-Southern Laos, the general offensive and uprising in the spring of 1968, the general strategic offensive of 1972, etc., to prepare a plan and execute the cryptographic task in order to ensure service for the command and control task in this great, historic campaign.

In order to fulfill in an outstanding manner the mission of service to campaign command and control, one of the decisive factors was the totality of the cadre and personnel of the army cryptographic branch, thoroughly permeated with the important significance of the general offensive and uprising, resolved in the strategy and clear-sighted ideological direction of the Politburo and the Central Military Committee.

Notes

1. Resolution of the Central Military Committee Conference of June 1973.

2. Extract from speech by Cde Le Duc Tho at the Vietnamese cryptographic branch conference recapitulating the cryptographic task in eight years of opposing America and saving the nation (21 March 1974).

3. Namely, Cdes Cam, Uong, Bau, Sinh, Chat, Thuc, and Khoa.

4. *Five Decisive Months*. Hanoi: People's Army Press, 1984, 182.

5. Notes of Cde Nguyen Van Khoi, cadre of the Bureau of Encrypting-Decrypting.

6. Holograph by Cde Vo Nguyen Giap, preserved in the General Staff Cryptographic Directorate

7. Cryptonyms of comrades Le Duc Tho, Van Tien Dzung, Pham Hung..

8. Group 232 was established in March 1975. [According to the *History of the PAVN* cited earlier, Group 232 was a corps-equivalent formation comprising the 5th and 3rd infantry divisions, reinforced by the 9th Division from IV Corps, plus branch troops. Commander, Maj. Gen. Le Duc Anh; political commissar, Brig. Gen. Le Van Tuong. Op. cit., 268, 271, 274. – Tr./Ed.]

Conclusion

The army cryptographic branch appeared immediately after the birth of our nation. As Uncle Ho said, "Cryptography must be secret, swift, and accurate." The army cryptographic branch strove to endure, to be worthy of the esteem of the Party. "In many decades past we fought bandits – we understood, and knew positively, that those were most important times, that our most secret problems could not be revealed."[1] In forty-five years of continuous building and fighting, under the command leadership of the Central Military Committee, the Ministry of National Defense, the General Staff, and of the echelons in the army, and the guidance of the Central Cryptographic Section, with respect to professional technical guidance, the army cryptographic branch was tempered in revolutionary struggle, and, with each passing day, grew up, "having taken part very significantly in the protection of secrecy to strengthen us to gain the victory."[2] Having come through two wars – against the French and against the Americans – the history of the building and fighting and maturing of the army cryptographic branch was tempered and found satisfactory in every respect, receiving from the Ministry of National Defense of our nation the golden words: "LOYAL, DEDICATED, UNITED, DISCIPLINED, CREATIVE."

With their mission and function--namely, to protect the nation's cryptographic secrecy, to ensure the secrecy of the contents of command leadership transmitted via cryptographic techniques – from the very outset, in revolutionary struggle as today in the building and protection of the Socialist Vietnamese Nation, the cadre and personnel of the army cryptographic branch have always carried deep in their hearts FIDELITY – to the nation, to the people, to the Party – prepared to struggle and to sacrifice for the combat objective and for the protection of secret matters involving the Party and the army.

From a very few cadre at the outset, entrusted by the army with the mission of performing the cryptographic task, the PAVN cryptographic branch built the ranks of professional technical cadre and personnel of the army's specialty branch with the quantity and quality to respond to the requirements of the revolutionary mission. These are the cadre and warriors who have been tested in combat service, in assignments all over the theaters of war, growing day by day in the level of the specialty and in absolute trust in the leadership of the Party, determined to safeguard and resolved to implement the victory policy and ideological viewpoint of the Party. These are the warriors of the PAVN, who embody the glorious essence, "Faithful to the Party, faithful to the nation, respectful of the people," whatever the situation, but also determined to fight and to serve and to swiftly transmit each message of command leadership from the various echelons of the army.

Cryptographic cadre and personnel are always firm in politics, enduring and overcoming every difficulty and test, always cultivating and training in revolutionary qualities and virtues, bound to and united with the organization, not one minute away from the combat objective of protecting secrecy, in order to fight and win out over the enemy in their silent battlefront. In the process of building and fighting, they produced countless examples of sacrifice of life and readiness to sacrifice life because of the mission – courageously fighting and wiping out the enemy, skillfully dealing with the concealment and destruction of cryptographic materials, or, when captured by the enemy, having to endure third degree treatment, yet refusing to talk and to disclose secret matters.

****

Dedication – that is the virtue prized highly in the tradition of those performing the cryptographic task. From new comrades entering the branch to comrades with decades in the branch, all of them hold high the [sense of] political responsibility, know to place value on the social class and the people [dan toc] above all else, before all else: cryptographic cadre and personnel gladly forsake personal tastes and wishes, voluntarily finding peace of mind in long-term service in the branch, well aware of being "unsung warriors" on the battlefront of keeping secrets; enduring hardships, overcoming difficulties, devoted and doing their best in research, creativity, and in building the branch in every respect. In routine times, as when going into battle; on mobile operations, as well as in emergency situations; days and nights of campaigning – work stacked up, requiring immediate attention – conditions in which materials were in short supply – illness, etc., army cryptographic cadre and personnel continued to be passionately absorbed in their work – forgetting to eat – forgetting to sleep, sticking to their studies, serving in combat, determined to ensure thereby that the arteries of command of the Party and of the army were fully in hand, secret and timely.

In the front lines as in the rear lines; in small, dispersed units as in large, concentrated units; in distant sea islands as on the mainland; in the enemy regions as in rear bases – wherever – insofar as the army cryptographic cadre and personnel, it was a silent battlefront, yet very urgent and not infrequently decisive. In order to have the right stuff to carry out the mission, army cryptographic cadre and personnel strove to study and train with respect to every aspect, with the slogan, "the amount of training produces skill; total absorption produces talent."

Industriously persevering, taking pains to ponder and seek out and bring into play innovativeness, improving technique to enhance the productivity and quality of the task – these are the manifestations of the dedication quite readily seen in each cryptographic cadre and warrior.

****

In obedience to the teachings of Uncle Ho, when he visited a cryptographic cadre and personnel development class in the Viet Bac combat sector [chien khu] during the war of resistance against the French, "The fellows doing cryptography must be secret and of one mind," the army cryptographic branch unceasingly turned its mind to building and consolidating unification. Unanimity with respect to ideological viewpoint and direction of the mission, as well as the guidelines and methods of the technical-professional task had changed into strength and unity of action from top down, so that, in whatever circumstances, the army cryptographic branch would realize victory in every mission.

In the branch – first and foremost at the basic level of cryptography, between cadre and personnel – not only deeply attached to one another with respect to ideals and revolutionary aims, but also deeply attached to one another in sentiment and collegiality – the cadre and personnel were like siblings, sharing the bitter and the sweet, sharing the good things in the daily specialist task, constantly creating favors for friends and keeping tough things for oneself. In the assignment or in ordinary life, the army cryptographic cadre and personnel constantly took care of the unified relationship to aid one another. This was the very beautiful style of the army cryptographic cadre and personnel.

Bringing into play the tradition of unity, the army cryptographic branch constantly stuck tightly to the leadership of the committees and commands at the various echelons, closely bound to the organizations in the command system at the various echelons, taking the initiative to build ties and close associations with friendly units, with the branches, and with allied organizations, so that together they could ensure the mission. Imbued with the international line of the Party and the teachings of Uncle Ho, that helping friends was helping oneself, with ardent love of country and pure international sentiment, our army cryptographers were fighting and giving their lives, shoulder-to-shoulder with their cryptographer friends in the Laotian army and Cambodian army: because of the revolutionary work of our friends, our army's cryptographic branch had helped in the building of the cryptographic branches of the Laotian and Cambodian armies, helping them to grow up and be self-sufficient in every respect.

****

Maintaining discipline is a feature of the army cryptographic branch that is fully satisfactory in every respect, an element in ensuring the trust of the ranks of cadre and personnel and of cryptographic technique, ensuring unity in thought and action, a strength of the whole branch. Perceiving clearly the characteristic nature of their task, the army cryptographers constantly settled on the importance of the problem of building standing operating procedures and implementing task discipline, in order to raise the sense of organization and discipline of the cryptographic cadre and personnel in every activity. The branch constantly educated and fostered in the cadre and personnel a sense of revolutionary vigilance and the spirit of discipline, paying particular attention to on-the-job workstyle and activities consistent with the nature of the mission of the cryptographic task: Voluntarily adhering strictly to the rules and regulations that had been set, not letting out secrets to anybody lacking the need-to-know; secretive and cautious in speech and work, in relations and connections, and struggling without compromise against manifestations of undisciplined freedom and lack of truthfulness and straightforwardness. Simultaneously striving to build and make routine the professional guidance and unity of administration throughout the army cryptographic branch, thereby creating a unity of all with respect to ideological awareness in content, line, and task methodology.

Faced with the requirements of the revolutionary mission, faced with the schemes and plots of the enemy, on the battleline of cryptographic secrecy, the army cryptographic branch proceeded to regularize and, step by step, to modernize the basic technical material, and the education, training, and bringing into play of tradition in order to raise vigilance, and the maintenance of discipline took on important significance. This was also a means of guarding oneself the best in resisting the enemy and ensuring that the building of the cryptographic branch was clean, strong and solid.

****

Firmly grasping and applying the Party's line of independence and sovereignty, the army cryptographic branch brought into play to a high degree the spirit of self-reliance, becoming stronger through one's own efforts, laboring to create and build the Branch, expressing the revolutionary spirit of attack, bringing into play the sense of socialist patriotism, along with the intelligence and creative capacity of the cadre and personnel of the entire branch.

As a secret, technical, scientific branch, carrying the characteristic traits of every distinct nation, thus having clearly defined sound ideology and having creativity in organization and application, we raised the spirit of "dare to think, dare to act," seeking to create, coordinate and apply the intellectual factors with the technical,scientific achievements, along with the experiences of real-world action over a long and arduous time in war and in peace, so as to research and create principles, methods, and forms of cryptographic technique and the art of organization and use to serve the army at a level that never ceased to move forward, struggling to defeat every means of signals intelligence [thutin ma tham] on the part of the enemy, thereby expanding science and technique.

****

Taking legitimate pride in the history of building and fighting while growing up, and the glorious tradition of the army's cryptographic branch in the past forty-five years, the army cryptographic cadre and personnel/warriors swear to hold fast and strive to bring into play their fine traditions, never ceasing to strive to train in revolutionary quality and level of output in every aspect, taking part in bringing about the victory resolution of the

186

standing Committee of the Central Military Affairs Party Committee with respect to the mission of building the army cryptographic branch in the new phase, to serve outstandingly in the mission of building and guarding the socialist Vietnamese nation.

Notes

1. From a speech by Cde Le Dzuan, General Secretary of the Central Committee of the Vietnamese Communist Party at the 1978 Nationwide Conference of Cryptographic Cadre.

2. Ibid.

Supplement

Cryptography in the Armed
Public Security Forces, 1959 – 1989

The Armed Public Security Forces – a sort of State police, border and coastal guard, and internal security force, distinguished from the PAVN by green, as opposed to red, collar tabs or shoulder boards – were formed in March 1959 in accordance with Politburo Decision 58 of November 1958 (HQ, Border Guard Troops, History of the Border Guard Troops, Vol 1, 1959-1979. Hanoi: People's Public Security Publishing House, 1990, 15). Shifting from its parent ministry (which had had its name changed to Internal Affairs) to the ministry of National Defense, and finally back to the Ministry of the Internal Affairs in subordination, they were initially (1959) issued a reserve PAVN cryptosystem to commence operations. In a 1989 publication (Socialist Republic of Viet-Nam, Armed Public Security Forces Staff, History of the Border Guard Cryptography, 1959–1989 (Draft). [Hanoi:]Border Guard Headquarters, 1989) appears an account that naturally complements the preceding history of PAVN cryptography and moves the account into the cryptomachine era. The following excerpts span the thirty-year period of APSF/border guard cryptography history to 1989.

****

"One of the most important tasks of the Forces' Cryptography at this time [i.e., at the outset, in 1959] was the urgent research and compilation of a code for the Armed Public Security [APS] in order to replace the DzB2 [Vietnamese DB2] code. The DzB2 code was an army cryptographic reserve [zu bi] type of code, supplied to APS Cryptography by the Cryptographic Bureau (General Staff) for temporary use in making contact during the time in which the Forces were newly established. The content of this type of code was not appropriate to the nature, mission, and sphere of activity of the APS; moreover, it was used all over the entire North – capacity was limited; timeliness could not be ensured.

"From mid-1959, the Cryptographic Section sent up to the Central Cryptographic Section a plan to compile a new code dictionary; at the same time, they instructed cryptography of the units to participate in a study of the frequency of words in messages. This approach was aimed at compiling code content consistent with the nature and activity of each unit, thereby producing high results in usage. . ." [p. 23]

"With high resolve and firm perseverance in the research, by February 1960 APS Cryptography had produced the first type of code for the Forces, designated 'Code T90'.

"The T90 code was constructed for superencipherment with cryptographic key having a reliable level of ensuring secrecy, belonging to the KTB4 type of technique. The use of

189

code T90 was very handy, easy to manipulate, productive in encryption, and twice as fast in decryption, compared with the DzB2 system.

"After the Central Cryptographic Section finished printing the T90 code by April 1960, APS Cryptography officially brought it into use throughout the liaison network of the Forces, and terminated the use of the DzB2 code." [23–24]

[In 1962, following Central Cryptographic Section directives to continue to improve cryptographic security, APS Cryptography coordinated with the Central Cryptographic Section to produce], "along the lines of code T90, codes type VQ1, VQ2, VQ3, VQ4, and VQ5 . . . and arranged to use them on the Vietnamese-Chinese and Vietnamese-Laotian border nets." [39]

"Effective from January 1966, the [APS] Cryptographic *Section* was elevated to become the Cryptographic *Bureau* (Decision 113/QD-CA, 18 January 1966)." [64n]

"[In 1968,] per instructions from the Ministry of Public Security, HQ, APS directed the Nghe An APS Command Section to cross over and coordinate with and assist our Laotian friends in that area of responsibility (the designator of the area of responsibility was K5) . . . . A cryptographic team comprising Master Sergeant Pham Xuan Hop and Dang Van Thong was ordered to move out with the command organization cadre group from the province, going over to the K5 area of responsibility. The cryptographic team took the designator DX5." [71] "A cryptographic team from the Ha Tinh APS (with the designator DX6) was active along with a provincial reconnaissance subunit in the Na Muong sector." [72]

CHANGING OVER TO THE USE OF THE NEW TECHNIQUE THROUGHOUT THE FORCES, RAISING THE LEVEL OF SECURITY OF COMMUNICATION CONTENT THROUGH CRYPTOGRAPHIC TECHNIQUE

"Schemes to destroy the Vietnamese revolution were carried out: The intelligence organizations, especially technical intelligence, concerned with the discovery of the contents of communications through cryptographic techniques, held the most important position. They said that 'information of the greatest importance, and latest and most reliable, is information obtained through the process of cryptanalysis.' Thus America's technical intelligence organizations had built a gigantic system for collecting information and cryptanalysis. This system was comprised of centers and bases positioned in many spots, all over the South, Laos, and Thailand, and on ships along the coast. They carried out deadly radio reconnaissance activities, using types of search-and-measurement equipments, pinpointing the locations of our transmitting and receiving stations, using transmitters, metal, etc., to jam, or for airplanes to shell and bomb. They collected our encrypted messages; they used the most modern and sophisticated equipment in order to search out the secret contents. Together with a large amount of modern equipment, hundreds of scientists and American professional technical specialists, the puppets [i.e.,

190

the forces of the Republic of Viet Nam in the South] were mobilized and used for the above objectives.

"In parallel with this activity, they sought to infiltrate into the cryptographic organizations to pilfer secrets, to buy out and win over or pressure cryptographic personnel to serve their ends.

"The more cunning the enemy's designs, the more we had to be vigilant, to raise the level of secrecy protection of our technique. The approach, 'mobilize to get a step ahead of the enemy,' in this arena immediately became the dominant direction of the Vietnamese cryptographic branch.

"Starting right out in 1964, the Central Cryptographic Section decided to carry out research on new technique.

"In 1966, the Central Party Secretariat issued Instruction No. 129/CT-TU (6 June) concerning the matter of 'Increasing the Preservation of Secrecy in the Radio Communication-Liaison Task of the Party and the Nation,' and the Prime Minister issued Instruction No. 96-TTg (6 June) defining 'Preservation of Secrecy in the Use of Telegrams.' [See above, p. 123]

"In 1967, after a time of consolidation, the ranks of scientific cadre researched and prepared. The Central Cryptographic Section decided that we must 'change over by positive steps to the use of technique KTB5 to replace KTB4 throughout the nation, in order to heighten the degree of secrecy protection of cryptographic technique. At the same time, expand the area of experimentation in KTC technique in a number of major liaison nets and make basic preparation to go into the total use of KTC.'

"At the end of 1967, thoroughly grasping the decision of the Central Cryptographic Section, the APS Cryptographic Bureau constructed a plan to implement the switchover to the new technique. The Party committee and commander of the Staff Directorate issued concrete guidance for each step of the implementation,with the requirements: positively, urgently, firmly by each step, changing to good technique means ensuring guidance and command requirements of the various echelons in every situation.

"At the beginning of 1968, the Staff Directorate opened a training class in technique KTB5 for cadre in charge of cryptography in the sectors, cities, provinces, and directly subordinate units, in order to thoroughly grasp the line of spreading the technique of the Vietnamese cryptographic branch and the requirements involved in the process of preparation and implementation of the changeover to the new technique.

"After training at HQ and returning, the units promptly organized short training classes for cryptographic cadre and personnel.

"In wartime conditions, with rank and file scattered over border posts, implementation of the plan encountered many difficulties. All of the provinces had to organize two to three training classes, with rotational replacements for those going to the provincial headquarters to study: they had to ensure both people for the continuing task and the plan's provision that 100 percent of the troop strength of the units would have completed study of the new technique. By October 1968, cryptography throughout the entire Forces had accomplished the training classes.

"At the same time it was guiding the units in implementing organization to develop the plan, the Cryptographic Bureau was researching and completing the compilation of the KTB type of cryptographic codebooks [lit., "dictionary codes," or "code dictionaries"], so as to have them so the nets could quickly put them into use.

"In December 1968, APS Cryptography in Thanh Hoa, Nghe An, Ha Tinh, Quang Binh, and Vinh Linh developed the use of the KTB5 technique comprehensively throughout their system of cryptographic technique. At the beginning of 1969, the development had expanded to the provinces of Quang Ninh, Hai Phong, Son La,and Lao Cai, and the internal net of the Viet Bac Sector.

"In October 1969, Lai Chau was the last unit of the Forces to change from the use of Technique KTB4. This was also the point in time that marked the conclusion of changing over to technique KTB5 in the entire cryptographic technique system of the APS.

"In order to promptly achieve productivity and quality in the use of KTB5 to attain the level of proficiency throughout the entire system of organization, Forces' Cryptography set up a 'burst of emulation study, study closely aligned with the real-world assignment'. After only a bit more than two months of being roused, many code clerks achieved high quality and productivity. Units with good study movements were Lai Chau, Nghe An, Ha Tinh, Vinh Linh, and the Encrypting-Decrypting Section of the Cryptographic Bureau.

"The changeover to the use of technique KTB5 marked a new stage in Vietnamese cryptographic technique. But not stopping there, the Central Cryptographic Section--after implementing research on technique KTC--decided to expand the experimental use of this type on a number of important liaison nets.

"In September 1969, the military regions and services in the North, Military Region Tri-Thien, Military Region 5, and HQ of the South began to test the use of KTC in contact with the Cryptographic Directorate of the General Staff. The APSF cryptographic system [he thong] was one of those under Central Cryptographic Section guidance in the experimental implementation of the use at a number of points. Preparing for this plan, at the end of 1967, APS Crypto assigned four comrades, Nguyen Van Mui, Ngo Quoc Bo, Nguyen Van Ba, and Nguyen Quoc An, to go study the use of technique KTC at the Central Cryptographic School.

"The Central Cryptographic Section handed over to APS Crypto KTC-type codebooks to research and compile, then to send up to the Section for printing--the Section would supply the types of crypto key.

"In accordance with guidance from the Central Cryptographic Section, Forces' Cryptography was to try out contact with KTC3 first, concentrated in three nets--Sector 4, Sector Viet Bac, and the coastal net. Based on the real-world situation, the Section sent along guidance: vis-a-vis the APS cryptographic system, immediately thereafter, change over to the KTC5 technique and prepare at once the rank and file of cadre and personnel to accept the new technique.

"Changing to the use of KTC5 posed numerous difficulties, requiring simultaneous resolution between people and technique. At that time the rank and file of Forces' cryptographic cadre and personnel had many comrades of advanced age, physically weak, with a cultural level not past second grade. This was the contradiction between the real-world capacity and the requirements of the technique.

"On 1 October 1970, the Staff Directorate sent a report from Comrade Minister Tran Quoc Hoan and the HQ concerning the situation of the Forces' cryptographic organization and cryptographic technique to solicit opinions and guidance. The minister instructed: 'We need to prepare a cryptographic force to guarantee the political standard, be in good health, have a third-level cultural standard--tenth is best--then organize to use KTC5'.

"Implementing the instruction from the comrade minister, on 13 November 1970 HQ issued guidance to the related directorates subordinate to the organizations of HQ and the provincial command sections: 'We must reexamine the number of cadre and warriors performing the cryptographic task, aiming at selecting comrades with sufficient standards consistent with conditions of employment of the new technique. Besides the matter of trustworthiness with respect to politics, they need to be in good health, have long service, have a cultural level of class six up--these numbers to be selected to go for refresher in the new technique for use on liaison nets from sector, city, and province with HQ. Afterward, we want to directly augment culture to the third level in order to build better conditions for the use of the new technique. In addition, those comrades lacking good health for long service need research to shift to other assignments in the Forces. We need to enroll sufficient numbers for training in the use of the new technique, with the standard being, Party members of the Vietnamese Lao Dong Party, from the working class, guaranteed to be trustworthy politically, in good health, with a cultural level three'.[1]

"Implementing the instruction, the Staff Directorate had a concrete plan to develop the related-task aspects for the entire Forces, in order to attain the requirement of changing technique.

"In March 1971, twenty-one unit cryptography comrades were selected for refresher in KTC5 at the Central Cryptographic School. After these comrades had studied and returned, the Cryptographic Bureau opened two supplemental classes in the use of KTC5 for seventy-three cadre and personnel (the first class, three months; the second class, six months). Along with instruction at the school and consolidated supplemental [training], a number of units had in-place supplemental training themselves for eleven other comrades.

"Tied right in with the organization task, the Cryptographic Bureau, as a matter of urgency, researched and compiled KTC5-type code books in accordance with guidance from the Central Cryptographic Section. A team of technique research cadre from the Bureau applied the methods, compared the sample code, and picked out from secret messages the vocabulary content of the Forces during this time, while at the same time they guided those in charge of cryptography in the units in becoming directly involved in 'frequency counting'. As a result of close guidance, cryptography in the units made concrete plans, set norms, and made allocations for each of their cadre and personnel to go deeply into research, to collect, and to sort out findings.

"After a year of diligent implementation, up to the end of 1970, the units belonging to the former Sector 4, Haiphong, Quang Ninh, Lai Chau, Son La, and Viet Bac Sector HQ in turn sent the Cryptographic Bureau thousands of 'plain elements' of high frequency from the secret message vocabulary of each unit.

"On the basis of contributions from unit cadre and personnel, the Bureau research team proceeded to analyze, select, and compile seven types of KTC5 codebooks--seventy sets of the least, 120 sets of the greatest. This quantity was sufficient to replace the entirety of the KTB5 code and set aside a reserve, against the prospect of having to replace a code while in use.

"In December 1971, after receiving more than five tons of professional means and the KTC5 cryptomaterials from the Central Cryptographic Section, the Cryptographic Bureau completed allocation to the liaison nets between HQ and the sectors, cities, and provinces.

"In order to ensure tight control from the very outset, the Cryptographic Bureau directed test contact over the net between HQ and the sectors, cities, and provinces. The Message Encryption-Decryption Section (Cryptographic Bureau) picked places for experimental guidance, and publicized to all units throughout the Forces the experience they derived." [75-81]

By 20 July 1972 the net from HQ with the sectors, cities, and provinces used entirely KTC5. [82]

By August 1972, all border posts, coastal defense, islands. [Ibid.]

On 30 October 1972, all were using KTC5 – KTB5 was out. [Ibid.]

"In parallel with the building of organization and service to the steerage and command of the forces, following the aim of developing the cryptographic technique of the security branch's cryptography and the aim of building the cryptographic technique of the APS, at this time it was clearly stated: 'Research into the improvement of technique goes hand in hand with change to new equipment, by steps introducing cipher machine technique into the service of the Forces' steerage and command, aiming at increasing the speed of cryptographic transmission'.[2]

"The realization of this objective had fundamental benefits. At the end of 1977/beginning of 1978, the Cryptographic Directorate of the Ministry of Internal Affairs [the Ministry of Public Security having become the Ministry of Internal Affairs in 1976] issued for APS, twenty-three sets of M111 cipher machines and two sets of 1Bautomatic teleprinter cipher machines [bo may ma truyen chu tu dong].

"This was the beginning of the development of modern technical equipment in the cryptographic technique system of the APS, creating step by step the capacity to use mechanical means of encrypting and decrypting at the primary CPs of the provinces and cities throughout the entire Force.

"The introduction of cipher machines into use in the units brought practical results, cutting short the time required for transmitting information, reducing the mental labor for the cadre and personnel directly involved in the use of the technique.

"The development of technique was closely connected to the development of the ranks of cadre to administer with a corresponding degree of professional steerage, to develop symmetry and synchronicity between man and technique – this was a long-term, fundamental measure, with respect to the Forces' cryptography. Therefore, immediately in 1977, the Forces assigned five cadre to go and perform the cryptographic task in the units and to be developed in the schools of higher education; at the same time, enrolling twenty-eight other comrades for development in a short course at the Border Guard Officers' School (Son Tay). Once their studies were finished in the general curriculum of the Officers' School, they were transferred to the 12th Battalion (at Xuan Mai, Ha Son Binh), where they further studied the content of the administrative task, guidance in the use of cryptographic technique by the APS cryptographic at provincial and city level, and basic knowledge with respect to the science of modern cryptographic technique." [103–104]

[Tracing the years, 1975-1979, with two wars, one in Cambodia and one with China, both of which involved the APSF:] "With 122 cryptographic organizations and 295 cadre and personnel (1974), by 1979 that had grown to 278 organizations and more than 500 cadre and personnel. The system of cryptographic technique [had gone from] using entirely manual methods until, by 1979, one third of the cryptographic organizations at provincial and municipal level were equipped with cipher machines." [145]

[10 October 1979, APSF was transferred from the Ministry of Internal Affairs (formerly the Ministry of Public Security) to the Ministry of National Defense and renamed the Border Guard Troops.] [148]

"The total number of cryptographic [units] as border guard troops, army-wide, was 280 units with cryptographic organizations, including two HQ CPs; thirty-one provinces and cities; thirteen regiments and regimental equivalents; twenty-four subsectors, battalions, and mobile companies subordinate to the provinces; and 186 border posts." . . . "The number of cryptographic cadre and personnel in the entire force at that time was 647 people." . . . "Vis-a-vis the system of technique: the various types of cryptographic materials, cipher machine equipment,and professional means were also transferred along with the organizational system." [149]

"As of May 1987, the synchronous teleprinter line [or circuit: duong tryen chu dong bo] between cryptography's cipher machines (TN-75) with the communications van (R-140-M) was officially active at the two HQ CPs (in Hanoi) and (Ho Chi Minh City), assisting in resolving in a fundamental way the volume of secret communications transmitted between these two CPs." [164]

". . . On 3 August 1988,. the two ministries, National Defense and Internal Affairs, proceeded to transfer the Border Guard Troops from the Ministry of National Defense [back] to the Ministry of Internal Affairs." [Ibid.]

"Along with the transferal of the organization of the forces, on 25 August 1988 the Army Cryptographic Directorate officially transferred the system of cryptographic organization and the system of cryptographic technique of the Border Guard Troops over to the Cryptographic Directorate of the Ministry of Internal Affairs for administration and professional guidance." [Ibid.]

[On 3 March 1989, the Border Guard Troops celebrated their thirtieth anniversary, 3 March 1959– 3March 1989.]

Notes

1. Instruction No. 24/CT-CY, 13 November 1970, signed by Brigadier General Pham Kiet, Commander and Political Commissar of the Forces.

2. "Aims for the APS Cryptographic Task, 1977."

www.ingramcontent.com/pod-product-compliance
Lightning Source LLC
Chambersburg PA
CBHW080528090426
42733CB00015B/2519